Designer's Guide to Building Construction and Systems for Residential and Commercial Structures

Designer's Guide to Building Construction and Systems for Residential and Commercial Structures

Treena Crochet

Allied Member ASID

Illustrated by

David Vleck

Prentice Hall

Boston   Columbus   Indianapolis   New York   San Francisco   Upper Saddle River

Amsterdam   Cape Town   Dubai   London   Madrid   Milan   Munich   Paris   Montréal   Toronto

Delhi   Mexico City   São Paulo   Sydney   Hong Kong   Seoul   Singapore   Taipei   Tokyo

Editorial Director: Vern Anthony
Acquisitions Editor: Sara Eilert
Editorial Assistant: Doug Greive
Director of Marketing: David Gesell
Senior Marketing Coordinator: Alicia Wozniak
Marketing Manager: Harper Coles
Marketing Assistant: Les Roberts
Associate Managing Editor: Alex Wolf
Project Manager: Alicia Ritchey
Operations Specialist: Deidra Skahill
Interior Designer: Kathy Mrozek

Cover Art: Fotolia
Lead Media Project Manager: Karen Bretz
Media Editor: Michelle Churma
Full-Service Project Management: Denise Botelho, Element LLC
Composition: Element LLC
Printer/Binder: Courier Kendallville
Cover Printer: Lehigh-Phoenix Color
Text Font: Frutiger

Credits and acknowledgments borrowed from other sources and reproduced, with permission, in this textbook appear on appropriate page within text.

Library of Congress Cataloging-in-Publication Data

Crochet, Treena.
  Designer's guide to building construction and systems for residential and commercial structures / Treena Crochet ; illustrated by David Vleck. — 1st ed.
    p. cm.
  Includes index.
  ISBN 978-0-13-241428-9
  1. Buildings–Environmental engineering. 2. Buildings–Mechanical equipment–Design and construction. 3. Buildings—Electric equipment–Design and construction. I. Vleck, David. II. Title.

TH6014.C76 2012
690–dc22

2011004922

10 9 8 7 6 5 4 3 2 1

Prentice Hall
is an imprint of

www.pearsonhighered.com

ISBN 10: 0-13-241428-7
ISBN 13: 978-0-13-241428-9

For my nephews, Kris and Travis.

CONTENTS

Section 2 COMMERCIAL CONSTRUCTION AND SYSTEMS 67

It is widely accepted that there is no better experience than on-the-job training, but in the practice of interior design, that philosophy could have dangerous consequences. The first construction project I took on was the remodel of a townhouse on the Intracoastal Waterway in south Florida. The owners were "empty nesters" and purchased a 20-year-old townhouse with three bedrooms upstairs, and an open kitchen, dining, and living room downstairs. They hired me to select new ceramic tile for the downstairs floors and new kitchen cabinets and countertops. Upstairs, they wanted to knock down the wall between the master bathroom and a third bedroom to make a larger master suite with an enlarged bathroom and closet area.

Common sense told me not to move too many plumbing pipes because my clients were conscientious about costs. The limitation, however, was that the couple wanted a separate whirlpool tub and shower enclosure to replace the all-in-one system they currently had. I drew up a plan for the upstairs changes and met with the contractor to discuss my ideas. We were upstairs in the current master bathroom and, as he reviewed my plan, he said, "This is great! We won't have to move the vent stack." I had no idea what he was talking about. Maintaining my composure, I looked him straight in the eye and said, "Terrific!"

When I left the job site I went home and immediately started poring through my books (these were the pre-Internet days) to try to find out what a vent stack was, which turned out to be a venting system that runs through the house and out the roof to allow hazardous sewer gas to reach the open air, and that maintains pressure for proper drainage. On this project, I was lucky; the remodeling of the bathroom could have been a disaster. From that moment forward, I realized that although you can't learn everything in school, an interior designer must stay on top of the game when entering the world of remodeling work and new construction.

When deciding to write this book, I was most interested in discussing the questions: What do interior designers need to know to get their designs built? What should they know about interior systems like heating and air-conditioning, plumbing, and electrical (even though they are not responsible for designing these systems)? Acquiring a basic knowledge of building construction and how the mechanical systems work, students of interior design have a better understanding of how their designs will work in the building envelope and how they will be affected by mechanical systems for both residential and commercial spaces.

Moreover, students and interior designers need to understand the structural components and limitations in a building and how these structural elements might influence their interior space planning. Although architects and engineers work out the details of structure and systems, respectively, interior designers must have a working knowledge of how various building systems affect the design of interior spaces, including heating, ventilation, and air-conditioning; acoustics; plumbing; and electrical systems; and the attributes and performance of specified building materials in conjunction with interior finish materials.

Designer's Guide to Building Construction and Systems for Residential and Commercial Structures is written with the student in mind. Using language that is directed toward students of interior design, you will learn about basic structural principles applied to the building environment through a review of common building methods, including timber framing, masonry, and steel construction for residential and commercial projects as applicable. You will also learn industry jargon pertinent to building construction, will be made aware of the work necessary to prepare construction documents, and will be exposed to building and accessibility codes through technical terms and vocabulary interspersed throughout the text. These terms are in italics and are defined in the glossary at the end of the book.

Structural systems covered in the text include foundations, beams and columns, floors, walls, roofs, doors, and windows. Mechanical and electrical systems, acoustical control, and plumbing are presented through an array of contemporary theories and techniques used in the design of buildings. Furthermore, code issues are presented in the context of the construction process. Codes are referred to in this book *as general guidelines only,* and students must be aware that code compliance begins at the municipal level. Whenever a new project is begun, the applicable codes in force *for the location of the building project* must be upheld. State and federal agencies may also impose codes beyond those supported through the International Code Council, which is the primary reference source for code issues discussed in this book.

Through photographs, detailed drawings, and case studies, students will gain insight to the construction methods used for built environments, beginning with the exterior building envelope and continuing through the interior finish details like millwork, floor coverings, walls, and ceilings. For the visual learner, photographs accompanying the text provide further explanations that show the application of building construction and systems. My goal in writing this book in a concise format with easy-to-understand text is to give students an easy guide to learn about building construction and systems in preparation for a successful practice as an interior designer.

DOWNLOAD INSTRUCTOR RESOURCES FROM THE INSTRUCTOR RESOURCE CENTER

To access supplementary materials online, instructors need to request an instructor access code. Go to www.pearsonhighered.com/irc to register for an instructor access code. Within 48 hours of registering, you will receive a confirming email including an instructor access code. Once you have received your code, locate your text in the online catalog and click on the Instructor Resources button on the left side of the catalog product page. Select a supplement, and a login page will appear. Once you have logged in, you can access instructor material for all Prentice Hall textbooks. If you have any difficulties accessing the site or downloading a supplement, please contact Customer Service at http://247.prenhall.com.

ACKNOWLEDGMENTS

This book is the result of many people who shaped my interest in construction methods and techniques throughout the years. I appreciate the time my father took to explain to me the various blueprints, with their strong smell of ammonia, that he brought home from work each evening from whatever project he was working on at the moment. For a little girl of seven or eight years old, it inspired me to design my own houses for my Barbie doll. And as the steak knives from mom's kitchen drawer came out along with rolls of Saranwrap, cardboard boxes soon became dream homes with all the bits and pieces of a real house, including operable doors and clear, transparent windows.

If it weren't for Rachel Pike, my department chair at Wentworth Institute of Technology in Boston, who gave me a great push when she scheduled me to teach the Building Construction and Systems class one week before classes began back in 1996, I would not be able to explain how things get built, installed, or finished. Knowing information and disseminating it to students is not always the easiest of tasks.

Despite the many challenges I experienced with clients and contractors while on seemingly endless construction sites, a million thanks go out for stretching the boundaries of my knowledge, design skills, and tolerance to new levels. If it weren't for Robert Smaldone, the general contractor with whom I worked on building a family compound in Kennebunkport, Maine, I would never have considered how important a book like this might be for clients as well as students to ease the pain of building a new home. He also proved that, yes, you can build a 10,000-square-foot home and have the clients moved in within a year's time.

Many thanks to James Cross, who supplied me with the numerous construction photos that appear in this book, and those generously offered through www.constructionphotography.com and Jeff Schaefer for Bob Moore Construction.

Last, but not least, without Vern Anthony, Editorial Director, at Pearson/Prentice Hall, the pages upon pages of notes I had prepared for the group of students at Wentworth would never have gone beyond classroom lectures. Thanks, Vern! Also, to the rest of the group at Pearson, my editor Sara Eilert, Alex Wolf, Doug Greive, and Alicia Ritchey, who kept the ball rolling throughout the laborious production process, thank you all.

I also appreciate the help of the following reviewers who helped to shape the content of this book: Margaret Jeffries, Pellissippi State Community College; Catherine Kendall, University of Tennessee at Chattanooga; JoAnn Wilson, Utah State University; Donna Weaverling Daley, The Art Institute of Philadelphia; Michael Dudek, Kansas State University; and Jeannie Ireland, Missouri State University.

Thank you all!

ABOUT THE AUTHOR

Treena Crochet is Associate Professor of Interior Design and Assistant Dean of the School of Architecture and Design at New York Institute of Technology (NYIT) in Bahrain. To the global campuses of NYIT, she brings more than 25 years of teaching experience to the classroom. Since 1994, Treena has consulted on residential design projects ranging from new construction to historic renovation and restoration interior work throughout New England. Her interior design work has been featured in numerous publications, including *The Boston Globe* and *Yankee Magazine*. An award-winning and bestselling author, she continues to write books on architecture and design for students and homeowners. For more information, view her website at www.TreenaCrochet.com.

A BRIEF HISTORY OF STRUCTURES, MATERIALS, AND METHODS

The greatest works of architecture throughout history are those that reflect innovation in structural design, materials used in construction, and engineering methods for each respective time period that pushed the limits of physics to build bigger, taller, and more daring buildings. These achievements in building and construction ensued from one of the most basic human needs: to provide shelter. Early nomadic humans living during the Paleolithic Era traveled on foot following migratory animals—the mainstay of their food supply. Throughout their travels, they weren't in the same place for too long and, where available in the region, caves provided the most convenient shelter; they could move right in without having to build. Moreover, huts dating back to 27,000 BCE, made from the bones of the woolly mammoth, reveal Paleolithic human attempts at building a protective shelter. The design of these huts formed an almost domelike structure. Large bones were stacked one row on top of another in a circular arrangement with makeshift *columns* supporting a curved roof (Figure I.1). The bones were then covered with grasses, hides, or mud to provide protection from weather. The hut could be dismantled and transported to the next encampment.

Portable shelter improved over time as tentlike structures appeared. Thin tree branches stripped bare of limbs and leaves were arranged to form a broad circle and were *splayed* toward the top, where they were tied together with either rope or animal tendon. The sloped branches and narrow apex stabilized the frame that became the structural support for a protective covering made of grasses or hides (Figure I.2). These building materials were readily available and it took the ingenuity of early humans to design a type of shelter that was sturdy and could be taken apart easily and carried with them on their hunt for food. The timber frames for these tents were lighter than carrying around a lot of woolly mammoth bones and much easier to transport from place to place. Today, tents or *yurts* are still used in

FIGURE 1.1 A drawing of a Paleolithic bone hut shows how animal bones were arranged to create a supporting structure for a hide cover that offered shelter to its inhabitants.

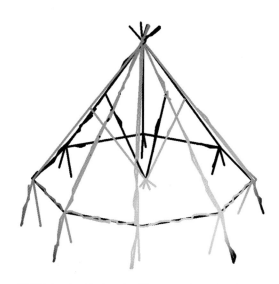

FIGURE 1.2 This drawing illustrates the timber framing for a tent, yurt, or teepee, which can be taken apart easily and transported from one settlement to another.

FIGURE 1.3 This yurt, seen in a village in Mongolia, is made with a willow structural frame and covered with rope-secured cloth.

some regions, like remote areas of Mongolia, and are experiencing a wave of new interest as weekend getaway houses in the United States (Figure I.3).

The resourcefulness of Neolithic settlers and their experimentation with the available materials surrounding them informed the designs of more permanent shelters. Grasses were woven, bundled, or braided together for roofs and wall coverings on huts, whereas bricks made from mud—mixed with straw and set in the sun to dry—provided structural building blocks. The process of stacking these bricks one on top of the other proved to be a relatively quick and easy method of construction (Figure I.4). Mud bricks were laid in several *courses,* and openings for doors and

FIGURE 1.4 The remnants of a mud brick wall still stands in Egypt as a testament to its strength and durability as a building material in hot, dry climates.

FIGURE 1.5 Huts found in a remote African village are made from mud brick and use thatch as a roofing material.

windows were reinforced by placing logs overhead as a structural *beam*. Lightweight *thatch* was used for covering the roof (Figure I.5). Mud brick houses were permanent structures and were best suited to those areas where the weather was dry.

In climates where rainfall came with the spring and summer seasons, stone proved to be a more suitable and long-lasting building material. Stone, quarried in close proximity to the settlements, was cut into uniform blocks and set in courses, one on top of the other, without the use of a *mortar bond*. The mere weight and size of each stone kept it in place and the walls strong. A thatched roof with small openings allowed smoke from the *hearth* to escape from the interior (Figure I.6).

Throughout the Neolithic period, quarrying rock for building projects became more aggressive, and new construction methods aided in building more impressive structures. The ruins at Stonehenge in England, with its familiar circular arrangement of large stones, date to around 2900 BCE and rely on a *trabeated* structural

FIGURE 1.6 Reconstructed huts at the Iron Age site of Citania de Briteiros in Portugal feature cylindrical stone walls and thatched roofs.

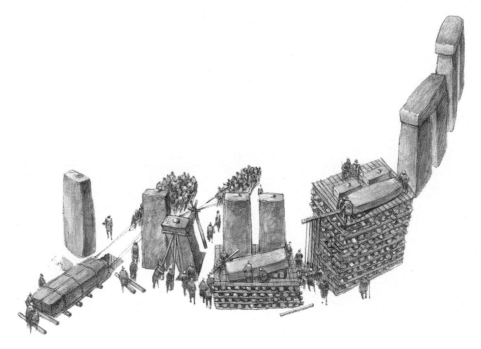

FIGURE 1.7 There are various methods used for raising massive megaliths into their upright position through leverage systems, and the type of scaffolding built for setting lintels in place during the building of Stonehenge.

system (Figure I.7). Trabeated structural systems are the most basic form of support, using a series of vertical *posts* to support horizontal *lintels*. At Stonehenge, large quarried stones called *megaliths* comprise the post and lintels. Like any tall structure today, the *foundation* of Stonehenge is the most important element in the construction process; without a sturdy foundation, the megaliths would not stand up.

The foundations for Stonehenge were formed by first mapping out the circular shape that would become the finished plan of the structure, then digging a ditch along the circumference. Workers relied on the physics of *leverage* by using a plank-and-fulcrum system (like a seesaw) to raise the megaliths into place (Figure I.8). When the posts were set upright, they dropped into the trench that then was filled in with aggregate and dirt to secure the post deep into the foundation. With the help of *scaffolding*, leverage, and *hoisting* mechanisms, lintels were positioned horizontally over the *span* of the posts. Because the wheel had not yet been invented in the British Isles, a primitive type of hoisting system using ropes and logs, along with muscle, lifted the lintels into place (Figure I.9).

In Egypt, the builders of the great pyramids at Giza combined the forces of workers and leverage to position 2½-ton blocks, one on top of the other, to form

FIGURE 1.8 The force of leverage to lift heavy objects acts in the same manner as a seesaw; exerted force at one end lifts weight at the opposite end.

FIGURE 1.9 Hoisting heavy objects with the use of a pulley system is easier by transferring the force of the person onto the rope. Pulling on the rope puts it in tension, and the continued exertion of force lifts the weight off the ground.

the structure. More than two million limestone blocks comprise the largest of the pyramids, the Great Pyramid of Khufu, which was constructed around 2600 BCE and reaches more than 450 feet into the air (Figure I.10). Rock was quarried and transported to the construction site from miles away. The quarried blocks were shipped on barges down the Nile River during flood season, when water levels were high, and were then brought to the Giza plain on rope-pulled sleds. When onsite, the blocks were cut to more precise measurements. Workers then used a

FIGURE 1.10 Most of the limestone facing material that gave the pyramids a smooth surface appearance has been stripped away to reveal the large sandstone blocks used to build them.

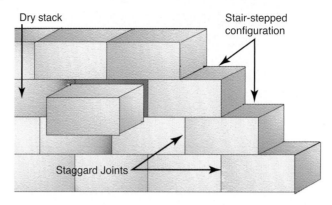

FIGURE 1.11 Egyptian workers stacked the large sandstone blocks in coursings, one on top of the other, over a laid foundation of limestone blocks. No mortar was used between the blocks; their weight was sufficient to keep them in place.

series of ramps and ropes to pull the blocks into position to form each new coursing (Figure I.11).

New Kingdom Egyptian temples like those in Karnak, built from 1408 to 1300 BCE, were designed as trabeated structural systems consisting of columns and lintels. A series of columns arranged in close proximity to the main *hypostyle hall* were 75 feet high and had diameters of 6½ feet. Perimeter columns and the inner columns of the hypostyle hall were massive, supporting the weight of a sandstone roof. However, their robust diameters significantly reduced the amount of usable interior space (Figure I.12). The huge stone columns were made of individual cut

FIGURE 1.12 A view of the tightly spaced columns in the hypostyle hall of the Temple of Karnak shows remains of the roof they support.

FIGURE 1.13 A large mechanical hoisting device was used to lift each section of a column into place during the building of the Temple at Karnak.

stone drums that were designed with interlocking joints. A large hoisting device similar to a *pulley* system was used to lift the drums, placing one on top of the other to create the column (Figure I.13). Interlocking joints between each drum stabilized the columns.

Greek temples built in the Ancient world followed similar construction methods as used by the Egyptians, but the Greeks made one significant improvement—the diameter of interior columns was greatly reduced to yield more usable interior space (Figure I.14). Perhaps the Greeks discovered that Egyptian temples were overengineered with their large-diameter columns. Greek temples had perimeter columns that carried most of the structural *load* of the roof with the help of strategically placed interior columns and interior *load-bearing* walls. Lintels spanning the columns, and load-bearing walls, supported timber *rafters* to form the shape

FIGURE 1.14 The interlocking joints used to stabilize each column section by locking it into place to the section underneath is seen in this collection of columns toppled by an earthquake.

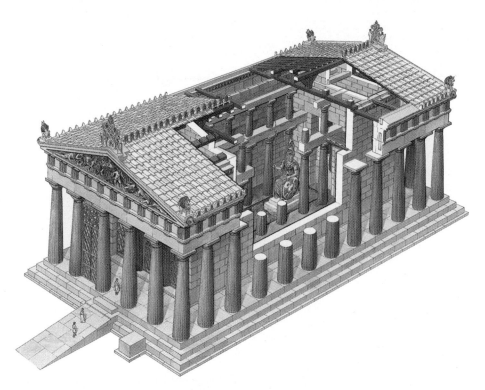

FIGURE 1.15 A reconstruction drawing shows the interior structural system for the Greek Temple of Aphaia. Columns support spans of lintels, which in turn support timber framing for the roof.

of the roof (Figure I.15). Trabeated structural support systems continued to be used throughout the first half of the Ancient Roman period, proving effective in the design of temples, apartment buildings, and civic buildings (Figure I.16). The discovery of *concrete* and its use as a building material put Roman engineering in the forefront of building construction.

FIGURE 1.16 This relief dating from the around 100 CE documents the use of a large crane to build a Roman temple. The great wooden crane was powered by a multitude of men working the wheel like a giant treadmill.

Roman builders discovered that by mixing volcanic ash, or *pozzolana*, with loose aggregate, water, and lime, the material dried to a rock-hard substance suitable for use as a building material. Pozzolana acted as a binding agent similar to *cement*. By using poured-in-place concrete, the Romans perfected an *arcuated* architectural support system that enabled the construction of large-scale building projects and domed structures. Remains of domed buildings, arcuated aqueducts, and massive amphitheaters seen throughout the former Roman Empire are a testament to their engineering achievements. Arcuated architecture relies on a support system of arches, *vaults*, and load-bearing walls, which the Romans stabilized with concrete (Figure I.17).

(a)

(b)

FIGURE 1.17 (A, B) An extruded arch or barrel vault acts as a directional passageway inside amphitheaters throughout the Roman Empire.

FIGURE 1.18 A section of the exterior wall of the Colosseum in Rome reveals a support system based on three tiers of arcades rather than load-bearing walls.

The Colosseum in Rome, finished around 80 CE, is an example of arcuated architecture (Figure I.18). Temporary wooden *formwork* supported the weight of stone blocks until the uppermost block, called a *keystone,* was inserted and the concrete mortar hardened (Figure I.19). When the arch was stabilized, the formwork was moved to create the next arch. In the Colosseum, each arch carries the building load from above, distributing it down around the arch and onto the foundation. The walls were built to a height exceeding 150 feet with successive arcaded tiers. Stone walls and piers were strengthened with concrete or *rubble stone* infill, and sturdy concrete foundations bore the weight of the structure (Figure I.20).

The Pantheon, built in 126 CE, became the Romans' crowning achievement in concrete construction. The cylindrical walls were formed by a series of arches filled in with brick, and the immense structure was topped by a poured concrete dome. The massive dome appears saucer shaped from the outside, but it is a perfect semi-sphere on the inside (Figure I.21). After the 20-foot-thick circular base was finished, formwork was arranged along the top, and concrete was poured in place one tier at

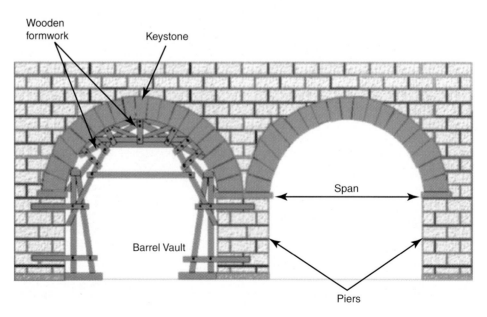

FIGURE 1.19 Wooden formwork was used as a temporary support structure for building arches out of stone blocks.

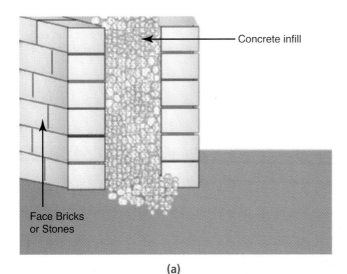

Concrete infill

Face Bricks
or Stones

(a)

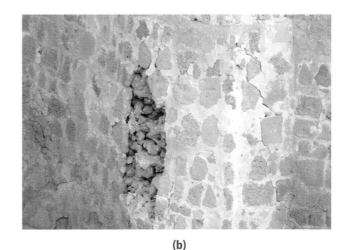

(b)

FIGURE 1.20 (A, B) Concrete and rubble stone infill was used to stabilize the lower walls of the Colosseum.

a time. Each tier was made thinner than the one beneath it until all tiers reached the top, leaving a 29-foot-diameter opening called an *oculus*. Impressively, the dome stands as originally built nearly 2,000 years ago.

By the Medieval period, builders minimized the use of concrete and reverted back to using stone blocks as a primary building material. The towering cathedrals designed in the Romanesque and Gothic styles utilized arcuated systems of structural support dependent upon the extensive use of cross-barrel vaults (Figure I.22). These vaulting systems supported the roof over the inner *nave* and side aisles to carry substantial structural loads, as walls were built higher than ever before. As the heights of the nave walls pushed toward the heavens, the vaulting system on its own proved insufficient. In 1284, in the cathedral at Beauvais, France, as winds pressed against the outside structure, the 157-foot-high walls became unstable and collapsed. Upon rebuilding, additional supports were added to keep the walls of the nave from crashing inward. External *buttressing* was attached to the exterior walls and distributed the weight down into the foundation (Figure I.23).

A quick glance at the campanile in Pisa, Italy, reveals medieval builders learned more lessons about structural stability. A reliable foundation system is an integral component in the structural dynamics of any building project. The architect never intended to design the "Leaning Tower" of Pisa; rather, he was unaware that the

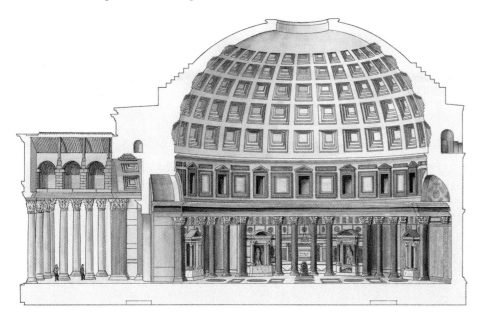

FIGURE 1.21 This section drawing of the Pantheon in Rome shows the diminishing thickness used in the construction of the concrete dome.

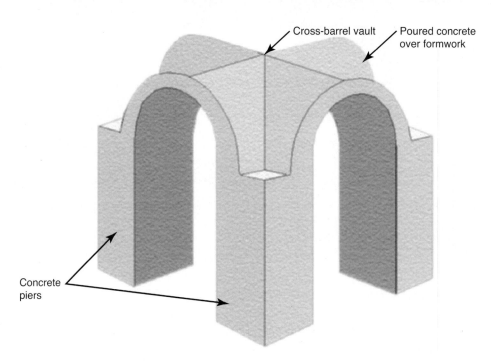

Cross-barrel vault

Poured concrete over formwork

Concrete piers

FIGURE 1.22 This diagram shows one bay of a cross-barrel vaulting system used to support the upper walls of medieval cathedrals.

soil where the church compound was sited was unstable (Figure I.24). The foundation for the bell tower was not built on solid rock; the area had once been under water. The tower began leaning during the construction process during the late 12th century, and throughout the tower's history, attempts have been made to stabilize the structure, which now leans nearly 15 feet off the vertical axis.

By the time of the Industrial Revolution beginning in the 18th century, new materials were introduced that transformed building construction. The development of large plates of glass (encouraged by England's repeal of the glass tax), and the manufacturing of iron *trusses* in the early to mid 1840s led the way for innovative designs. The construction of the Crystal Palace built in London in 1851 to house the Great Exhibition introduced interlocking iron trusses and floor-to-ceiling glass, setting the tone for fresh design (Figure I.25). The building materials used to construct

FIGURE 1.23 The north side of Notre Dame cathedral in Paris shows how external buttressing, often called *flying buttresses*, provides additional support to the wall structure.

FIGURE 1.24 The Leaning Tower of Pisa tilts precariously 14.8 feet off the vertical axis.

FIGURE 1.25 This interior view of the Crystal Place in London shows a structural support system of iron columns and trusses. The entire structure was enclosed in plate glass, a new invention at the time.

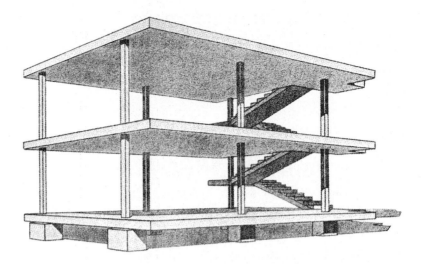

FIGURE 1.26 A drawing of Le Corbusier's concept behind the Dom-ino House of 1914.

the 770,000-square foot structure were entirely *prefabricated* then shipped to the site for assembly. The builders of the Crystal Palace used manual hoisting systems to lift the iron members into place, yet in 1852, manual methods were replaced with the invention of the portable steam crane. The iron trusses were fabricated in standardized sizes all based on a 4-foot module. This method of standardizing the size of structural components set a new paradigm for building construction that would be fully utilized by the architects of the early 20th century.

Swiss architect Charles-Édouard Jeanneret-Gris (commonly known as Le Corbusier) set the benchmark for implementing standardized sizes in building materials and using prefabrication in building technology through his concept of "unit" construction. His drawings for the Dom-ino House in 1914 defined the efficiency of the theory of architectural structural design based on a component module that could be duplicated and connected like a game of dominoes: up, down, and sideways (Figure I.26). The initial rectangular unit of the Dom-ino project was designed to be supported by six steel columns supporting an overhead concrete slab. The columns eliminated the need for interior load-bearing walls, leaving the interior space completely open. Because the structural load is carried by the perimeter columns, any material—like plate glass—could be used to enclose the structure. Le Corbusier's drawing for open-plan interiors reinforced contemporary concepts for reducing the structure to its most basic form: trabeated structural supports.

Further innovations in building technologies were realized in 1934 in the construction of the Kaufmann House designed by Frank Lloyd Wright. Nicknamed "Fallingwater," a *cantilevered* structural support system enabled the house to extend over a rushing brook (Figure I.27). The column piers supporting 15-foot cantilevered extensions were made from *reinforced concrete*, which was the primary building material for the remaining structure. Wright had pushed the structural limits of cantilevered systems and, against his objections, the client consulted a structural engineer who insisted on installing more steel reinforcing bars than originally specified by Wright. It was only during the mid 1990s that the cantilevers had *deflected* a full 7 inches off horizontal, and money was raised to stabilize them by adding *post-tension cabling*.

Post-tension cabling used in bridge construction, which transferred loads laterally to support larger spans, inspired architects to use these same structural concepts in their designs. Architect Eero Saarinen and structural engineer Joseph Vellozzi worked out the logistics of using post-tension cabling to build Dulles Airport in Virginia in 1962 (Figure I.28). Steel cables stretching 141 feet from one exterior wall to the other were set in tension to keep the outward-leaning walls in place. The steel cabling supported a sweeping roof comprised of interlocking concrete panels.

Had it not been for exploring alternative means of structural support mechanisms in the 1960s, the exceptional work of modern architect Frank Gehry would not be possible. Post-tension cables and the development of the tower crane during

FIGURE 1.27 Fallingwater, designed by Frank Lloyd Wright in 1934, utilizes perilously placed cantilevers to extend over a brook.

the late 1950s and early 1960s gave the architects of this generation new tools for designing sculptural structures that challenge the laws of physics and engineering. Studying architecture during the late 1950s no doubt shaped Gehry's notoriety for experimentation with daring structural forms, establishing him as one of the leading architects of the 21st century. At the dawn of the new century, his design for the Guggenheim Museum in Spain realized new engineering achievements at the time (Figure I.29). The integrity of the building's structural support relies on calculated engineering, rigid cores, and structural columns tied to post-tension cables and slabs to achieve mass that appears to float through cantilevered extensions. Structural loads formerly carried laterally by rigid *I beams* were replaced with support beams twisted into slender undulating rods that hold up sweeping canopies and misaligned exterior stainless steel panels.

FIGURE 1.28 Graceful curved rooflines and outward-angled walls seem to defy the principles of gravity in the design of Dulles Airport by Eero Saarinen and introduces concepts taken from the structural engineering of suspended bridges.

FIGURE 1.29 The Guggenheim Museum in Bilbao, Spain, designed by Frank Gehry in 1997, captures advanced skills in structural engineering enabled through the use of computer technologies.

Which new technologies or materials will shape the architecture of the 21st century remains to be seen. As long as designers imagine new and innovative designs, there will be structural engineers who will figure out how the buildings will stand. It is hard to imagine that only 50 years ago, the cantilevered designs of Frank Lloyd Wright were questioned as if the laws of physics would undoubtedly cause the collapse of the building. Perhaps since landing humans on the moon in 1969, generations since then realize anything is possible.

Residential Construction and Systems

Residential Construction
Building the Envelope

Homebuilding Basics

What goes in to building a new home? It's not just the style of a home that is an important consideration for builders and homeowners, but also (and more important) how the house will be built from the ground up. Will the builder and homebuyer agree to use *sustainable* materials or *reengineered* materials while considering *green design*? How will the performance of appliances and lighting, along with heating and air-conditioning systems going in to the new home stand up to the energy standards established by the U.S. federal government? Moreover, what building codes apply to residential construction projects? This section on "building the envelope" addresses these questions while explaining the basic principles of the construction process. Understanding how a home gets built helps the interior designer work with clients—whether the clients are building a new home, *renovating* an older home by remodeling or constructing a new addition, or *restoring* a vintage home.

Today, most new homes constructed in the United States usually fall into one of two categories: (1) those commissioned to be built by the potential homeowner

This new home, with its traditional styling, gabled roof covered in asphalt shingles, dormer windows, and covered entry, conveys the American dream of homeownership.

and (2) those built for speculative purposes. The future homeowner purchases the land, usually hires an architect to design the home to his or her specific purpose, and also employs a building contractor to do the work (often a general contractor recommended by the architect). Spec houses—those built on the speculation that someone will buy them—may be built by individuals or property developers who purchase land for a complete subdivision or for just one home. Regardless, these homes go through similar processes of securing the architectural drawings and hiring a general contractor to oversee all phases of construction.

Whenever a future homeowner commissions an architect to design a home, the design is generally a collaborative effort. It may take several meetings between the client and the architect to develop a set of plans that satisfies the client's needs and meets all requirements detailed in current *building codes*. Sometimes, potential homeowners might purchase *stock plans*—a prepackaged set of drawings complete with detailed construction documents for the contractor to follow. Often, during the initial stages of the design process, the client may hire an interior designer to review the preliminary architectural plans and to offer suggestions for interior planning. The interior designer may also design built-ins like bookcases, entertainment units, wet bars, and window seats; and may review the electrical plans to check the locations of lights and *switches*, and to help determine the stylistic feel of the interior. Furthermore, the interior designer and client formulate the desired aesthetic and select plumbing and lighting fixtures, paint colors, flooring materials, bathroom and kitchen tile, countertop materials, and cabinetry.

CAUTION

Housing Types and Building Codes

Single-family dwellings, duplexes, townhomes, apartment buildings, and condominiums are typical residential housing units in the United States, and all must be constructed to meet building code requirements. All codes are intended to protect the health, safety, and welfare of all occupants or users of the building. These regulations vary from state to state, and even from town to town. It is the job of the architect and general contractor to know the codes that need to be addressed and followed during design and construction. Code regulations for single-family dwellings are the least restrictive compared with those required for apartments and condominiums.

Codes are organized around occupancy classifications or use groups relative to the intended habitation of the building. The Residential Use Group includes all structures in which individuals live or in which sleeping accommodations are provided (with or without dining facilities), excluding those classified as institutional occupancies. Within the Residential Use Group, further subdivisions clarify the specific type of project:

R1 Occupants are primarily transient in nature. This includes boarding houses, hotels, and motels where the stay is less than 30 days.

R2 Multiple dwellings in which the occupants are primarily permanent in nature. This includes apartment houses, convents, fraternity and sorority houses, monasteries, rectories, rooming houses where occupants are not transitory, and dormitories that accommodate six or more people.

R3 Occupancies primarily permanent in nature and where building units do not contain more than two dwelling units. This subgroup also includes adult and childcare facilities for as many as five persons for less than 24 hours at a time.

In this chapter, code references are presented for the R3 occupancy use group. Some of the more restrictive codes for this use group are as follows: operable windows must be in all sleeping rooms, bathrooms must have an operable window or have installed a mechanical ventilation system, and the adjoining wall between the garage and living space must be constructed to meet a one-hour fire rating. In addition, smoke detectors are required in living and sleeping areas.

Cities, counties, and/or states may impose stricter codes or additional codes pertaining to areas prone to hurricanes, earthquakes, and flooding. With any project, it is important to check with the municipality where the project will be built to determine the codes that must be followed.

The International Code Council is an internationally recognized association and, according to the organization's mission statement, provides "the highest quality codes, standards, products, and services for all concerned with the safety and performance of the built environment." The International Code Council is the model code agency for which most codes are adopted. For further information, see www.iccsafe.org.

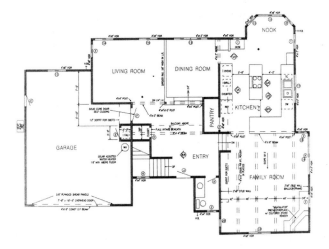

FIGURE 1.1 A set of construction documents includes floor plans, building elevations and sections, and other pertinent drawings needed to convey details for construction.

For residential construction, an architect prepares a set of construction documents for the building contractor to follow to complete the building process. In most cities or towns, these documents are first reviewed by a city official to ensure there are no building code violations. Construction begins when the plans are approved and a *building permit* is issued. The criteria for what must be included in a complete set of plans for code review are determined by the local municipality. A set of *construction documents* usually includes foundation and framing plans, and *floor plans* that detail the location of exterior and interior walls, doors and windows, plumbing fixtures and appliances (for determining electricity needs), heating and air-conditioning equipment (including hot-water heaters), and built-ins (Figure 1.1). The documents also must include an electrical plan that shows locations of electrical outlets and related electrical equipment (such as lighting, switches, ceiling fans, and hardwired smoke detectors).

Structural drawings that detail the framing of the house, including structural walls, floor and ceiling *joists*, and roof rafters with supporting beams and *trusses*, are also part of the set of drawings. Elevation drawings of the exterior, along with interior elevations to show construction details for built-ins, kitchen cabinet configurations, and wall tile patterns in bathrooms (accompanied by plumbing fixture schedules) are also included in the set of construction drawings. Although these are the most basic set of drawings that make up a construction document package, other drawings may be necessary, depending on the type of construction (*timber frame* or *masonry*) and geographic location. Homes in hurricane- or earthquake-prone areas require special construction considerations to ensure safety during natural disasters.

Regional Considerations

Along with areas prone to natural disasters, other factors like climate conditions (such as heavy snowfall or hot, humid weather) affect how a house is designed and built. Seasonal weather changes determine the materials used in construction, the orientation of the house on the lot, the type of windows featured, and the roof design. Positioning the house according to its exposure to the sun plays a significant role in *heat gain* or *loss*, which may impact the performance of heating and air-conditioning systems. *Passive solar* design maximizes the orientation of the house to the sun, resulting in more efficient heating or cooling. Building with regional materials chosen for their thermal properties like brick, clapboard, cedar shakes, concrete block, or adobe helps moderate building costs and makes the most of local resources.

In New England, where timber is plentiful, most houses are built with wood framing. Also, the steep gables of a traditional-style colonial enable the snow to slide off eaves, rather than build up, which then causes ice dams or collapse. Furthermore, limiting the placement or size of windows on the north side of the house helps keep out the cold during harsh winters, whereas south-facing windows take full advantage of heat from the sun.

In contrast, areas of coastal Florida rely on masonry walls to keep the house cool during hot and humid summers, and masonry construction provides a sturdy structure to stand up to hurricane-force winds. In these areas, roofs vary from sloped to flat, because snow loads are not a problem, and windows are coated with a film that keeps interiors cool by providing maximum protection from the sun's ultraviolet (UV) rays. In addition, high ceilings are common design elements in the main living space to take advantage of one of the laws of physics—warm air rises and cold air sinks—which helps the air-conditioning systems keep the interior cooler.

During the 19th century, pioneers spread westward across the plains of North America looking for fertile soil to start farming. For many, carving out an existence on the flatness of the midwestern plains wasn't easy, and these hardy pioneers quickly discovered that the plentiful earth surrounding them might be the best material for building a shelter. These early settlers discovered that small houses made from sod proved to be warm in the winter and cool in the summer, with an interior temperature of 50 to 60°F. Although these colonists were not the first to develop passive solar housing, they certainly contributed to its development in North America.

Passive solar design relies on exposure to the sun to keep a house cool during the summer and warm during the winter. These homes are sited on an east–west axis, with glass and other thermal conducting materials located toward the south, maximizing exposure to sunlight from around 10 o'clock in the morning to 3 o'clock in the afternoon. Depending on climatic variations, construction materials used on the southern side of the home are selected either to reflect or absorb the heat from the sun. Absorbed heat radiates into the home after the sun sets, whereas reflected heat keeps the interior cool. There are many factors to consider in passive solar design, including the tilt of the earth's axis relative to the sun during winter and summer months, and the use of overhangs to protect the interior from direct sunlight and heat gain during the summertime. For further information, consult the Northeast Sustainable Energy Association at www.nesea.org.

This photograph from the 19th century shows a pioneer family in front of their sod house built on the Nebraskan plains.

The south-facing wall of this home in Nebraska is made of solar collecting panels that supplement the home energy supply with electricity for running lights, heating and air-conditioning units, and a hot-water heater. Passive solar houses like this one benefit from sloped glazing, which absorbs more of the sun's heat.

Natural disasters like earthquakes pose another problem for building houses. In southern California, for example, foundations must withstand seismic changes caused by small tremors without causing significant structural damage to the home.

Foundation Systems

Starting with first things first, houses need a stable foundation on which to build. No structure can stand without a solid foundation; it is the most fundamental building component. The type of foundation for new-home construction is determined by the geographic location of the building site. Regional considerations at the building site include soil conditions, the presence of groundwater, and the depth of frost levels beneath the ground. A foundation plan shows the system that will be used and may range from poured concrete *footings*, a slab poured directly onto the

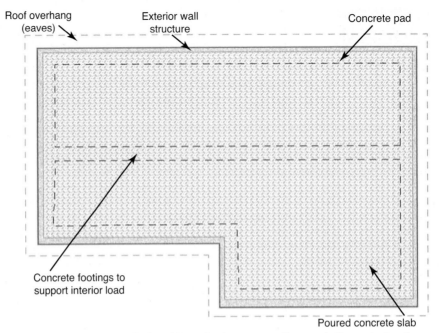

Roof overhang (eaves) Exterior wall structure Concrete pad

Concrete footings to support interior load

Poured concrete slab

FIGURE 1.2 This example of a foundation plan for a stem wall system indicates the locations of the perimeter and interior footings that will support the structural load of the home.

ground for hard soils and warm climates, masonry or treated wood pier-and-beam foundations, or basement foundations (Figure 1.2).

Soil conditions and regional climates influence the type of foundation system set in place. For example, a basement foundation in Texas is not practical, where rock-hard and clay-based soil leads to cracking and shifting, which causes damage to the below-grade foundation. In these areas, a slab-on-grade foundation works well. However, in the Northeast and Midwest, basement foundations are more commonplace. Basement foundation systems utilize masonry walls to shore up the excavated site and to support the main structure while offering the extra bonus of providing more interior space on a small plot of land. Pier-and-beam foundations made of masonry or treated wood piers to raise the house above the ground are typically used in areas where flooding or high water tables occur, such as in southern Louisiana or the Mississippi Delta. Understanding the different types of foundation systems is important in determining how the structural *load* of a house is supported.

Also, becoming familiar with the characteristics of each foundation system directs the interior designer toward making the right choices when specifying materials for the finished floor, considering both the weight of the materials and their suitability. For example, laying a hardwood floor on top of a concrete slab requires special preparation to avoid buckling; similarly, laying heavy stone or tile on a pier-and-beam foundation could lead to floor deflection, which would cause the tiles to crack.

LEARN More

Concrete vs. Cement

The ancient Romans discovered that by adding pozzolana, a locally abundant volcanic ash, to a mixture of lime, aggregate, and water, the material set to a stonelike hardness. Essentially, the pozzolana became the binding agent, or cement, that kept the other materials together to form concrete. The cement used today is made from calcium, silicon, aluminum, and iron ore found in limestone, sand, shells, chalk, shale, and clay. Concrete is a mixture of cement, sand, aggregate, and water. For further information, visit the Portland Cement Association's website at www.cement.org.

BASEMENT FOUNDATIONS

A basement foundation system begins with the excavation of the site dug deep enough to accommodate the finished ceiling height of the space. The thickness of the slab, ceiling joists, and finished ceiling material must be calculated to determine the proper basement depth. When constructed, the *below-grade* basement walls are sturdy enough to keep the dirt at bay and to support the weight of the house above (Figure 1.3). After the site is excavated, concrete footings are set on compacted soil and the basement walls are either poured-in-place concrete using formwork or they are built from stacked and mortared *concrete masonry units (CMUs)*. Both construction methods require steel rebar reinforcing. After the concrete sets and the formwork are removed, a 4-inch slab of concrete is poured over a gravel bed covered with a laid-in moisture barrier to prevent water from seeping into the finished basement. Basements designed to be entirely enclosed need windows positioned above grade, located in the upper section of the wall, because codes require light and ventilation in all habitable rooms. Depending on the slope of the site, it is possible to design a walkout basement with one wall fitted with a door and windows (Figure 1.4).

CRAWL SPACE AND PIER-AND-BEAM FOUNDATIONS

Crawl space and pier-and-beam foundations are used in moderate climates where water tables are high or there is a chance of flooding. Crawl space foundations have concrete footings and a short wall that supports the structural load of the house (Figure 1.5). The wall is raised above grade, creating a "crawl" space between the ground and floor joists. Access to the crawl space is usually through a hatch either inside the house or from a location in the perimeter wall (Figure 1.6).

In both crawl space and pier-and-beam foundations, the elevated floor avoids direct contact with the ground, which could become saturated with floodwaters or rising water tables. Furthermore, a moisture barrier such as Visquine, a brand name for plastic sheeting material marketed to building contractors, is laid on the ground to prevent moisture damage to the floor joists and the *subfloor* (Figure 1.7).

Like crawl space foundations, pier-and-beam-type systems also leave space between the floor joists and the ground. Pier-and-beam foundations are made from concrete, treated wood, brick, or stone and the piers are sized according to the weight loads they will carry. Piers and beams become the tie-in or anchor point for floor joists and they support a series of beams that distributes the load of the structure from pier to pier. Piers are positioned along the perimeter *footprint* of the house and the interior, and are placed strategically to hold up interior

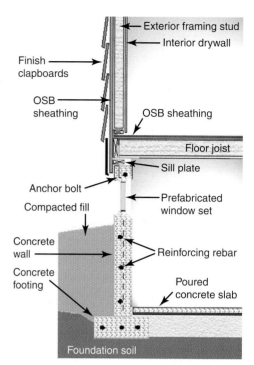

FIGURE 1.3 This section drawing details the component parts of a basement foundation system.

FIGURE 1.4 The walls for this walkout basement were made by pouring concrete into formwork made from sheets of plywood held in place with metal reinforcing bars. After the concrete set, the formwork was removed and a 4-inch-thick concrete floor was poured. A small window is visible on one of the poured-in-place concrete walls.

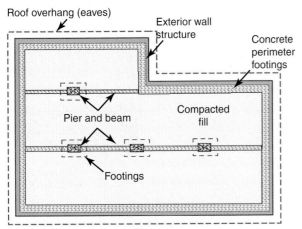

FIGURE 1.5 This sample plan for a crawl space foundation system shows the positioning of a perimeter wall with additional piers spaced throughout the center of the structure, which are designed to support the floor system and interior load-bearing walls.

FIGURE 1.6 Construction workers guide a prefabricated ranch-style home onto a crawl space foundation.

load-bearing walls needed to support either a second story or an attic space above (Figure 1.8). For coastal areas where reoccurring high tides or storm surges are prevalent, a pier-and-beam foundation can rise several feet off the ground, which keeps the house far above rising waters (Figure 1.9).

STEM WALL FOUNDATIONS

Stem wall foundations are commonly used in cold climates or areas with high water tables. Concrete footings are placed below the frost line (which varies in depth depending on geographic region), and a short foundation wall of either poured concrete or CMUs becomes the main support for the structure (Figure 1.10). A moisture barrier is laid on grade, and the space between the stem walls is filled with compacted soil, onto which a concrete slab is poured. Compacted soil in the stem wall acts as insulation during cold winter months.

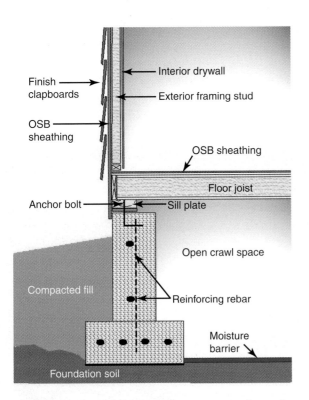

FIGURE 1.7 This drawing details the components of a crawl space foundation. A short masonry wall becomes the support for the home's exterior walls and raises the floor joists above grade. The space between the floor joists and ground level is kept open for access to the crawl space.

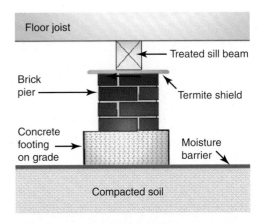

FIGURE 1.8 This detailed drawing shows the construction of a concrete footing and brick pier for a pier-and-beam foundation system.

FIGURE 1.9 A pier-and-beam foundation raises this beach house high above the ground, away from floodwater or tidal surges. The foundation piers are made from treated wood to resist rot from water or high humidity.

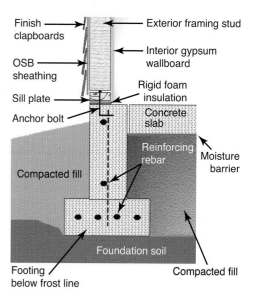

FIGURE 1.10 A section drawing shows the components of a stem wall and slab foundation system.

MONOLITHIC FOUNDATIONS

Monolithic foundations are the easiest to build and the most economical type of foundation system available. These monolithic pour foundations, commonly called *slab-on-grade foundations,* appear in warm climates where there is only slight chance of frost (Figure 1.11). Wooden formwork is laid out in sections according to the footprint of the house. The sections allow for easy leveling after the concrete is poured over a layer that consists of a protective moisture barrier and gravel. *Rebar* reinforces the concrete to stabilize the foundation against cracking and shifting (Figure 1.12).

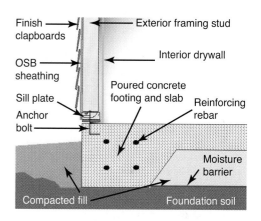

FIGURE 1.11 Slab-on-grade foundation systems are constructed by pouring concrete over a grid of rebar, which is laid on top of a moisture barrier in direct contact with the ground.

FIGURE 1.12 Concrete is poured into the formwork and then spread by construction workers when making a slab-on-grade foundation.

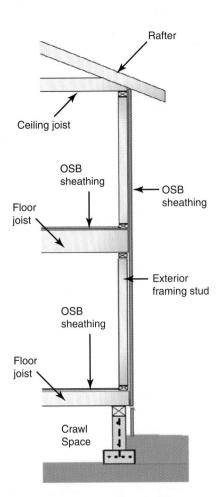

Rafter

Ceiling joist

OSB
sheathing

OSB
sheathing

Floor
joist

Exterior
framing stud

OSB
sheathing

Floor
joist

Crawl
Space

FIGURE 1.13 This section drawing shows the load-bearing components of a timber frame structure. The wood framing system includes two-by-six exterior studs, two-by-eight floor joists, and prefabricated roof trusses. The framing structure is supported by a pier-and-beam foundation system.

Wall Systems

WOOD FRAME CONSTRUCTION

Wood frame construction is widely used in the United States to form the structural framework for today's single-family homes and small apartment complexes. Nationally harvested pine, spruce, and fir are milled into framing *studs* in lengths of 8, 10, and 12 feet. These studs measure, in actual size, 1.5 inches by 3.5 inches, yet are referred to by the industry as *two-by-fours.* In addition to two-by-fours, two-by-six studs maybe used for the exterior framing on buildings that require a deeper wall cavity to accommodate thick insulation or large-diameter plumbing pipes and waste stacks. Vertically arranged wall studs are attached to a horizontally positioned wooden top plate and bottom sill plate for stability, and the wall section is then anchored to the foundation (Figure 1.13).

Studs used to frame exterior and interior walls are positioned between the top plate and bottom sill plate every 16 inches or 24 inches on center; usually, studs placed every 16 inches is more common. Wood frame exterior wall sections are framed out with double and sometimes triple studs around door and window openings to provide optimum strength for supporting these features (Figure 1.14). Completed exterior and interior wall framing, along with the first floor ceiling joists, act as a structural platform for framing the second floor or attic space above (Figure 1.15).

After all the framing is completed, *oriented strand board (OSB)* cladding is attached to the exterior of the studs (Figure 1.16). OSB is typically used instead of *plywood,* because this engineered material saves trees by using wood segments and adhesives. OSB creates a strong *sheathing* material for use on exterior walls, the top of roof rafters, and over floor joists. The cladding material is then covered with a polyethylene fiber paper that offers waterproofing and moderate thermal protection (Figure 1.17). Other cladding materials include fiber cement board, insulative board, and foam sheathing panels.

FIGURE 1.14 Timber frame wall sections are first constructed on the ground, and are then raised into place and anchor bolted to the foundation.

FIGURE 1.15 A second story is added to the completed framing of the first floor.

The sheathing is covered with a wide range of finishing materials, including vinyl siding, wood or engineered clapboards, natural cedar shakes, and brick or stone veneers chosen for their aesthetic appeal (Figures 1.18–1.20). Manufacturers sell reengineered materials made from recycled plastics, rubber, wood chips, or metal that provide attractive exterior cladding materials and protect the environment by saving our natural resources.

FIGURE 1.16 Oriented strand board (OSB) sheathing is applied to the stud walls as the enclosing skin. Large sheets of OSB enhance the thermal properties of the exterior wall system, leaving fewer seams for air to infiltrate the interior.

FIGURE 1.17 The OSB sheathing has been covered with Typar house wrap, a polyethylene paperlike sheeting with a waterproof backing that offers weather protection before the final finishing materials are applied.

FIGURE 1.18 Vinyl clapboard-style siding is applied as the finishing material on this new home. The reengineered siding is available in a variety of colors, it eliminates the need for painting, and the material can be recycled.

FIGURE 1.19 Cedar shakes cover the exterior of this New England home.

FIGURE 1.20 A home in the Southwest is finished in nonload-bearing brick veneer, which is essentially one thickness of brick laid over the exterior sheathing.

LEARN More

Habitat for Humanity International

Habitat for Humanity International (HFHI) was founded in 1976 by Millard and Linda Fuller as a nonprofit organization with a mission to provide housing for low-income families across the globe. Working through a network of affiliates, HFHI builds and rehabilitates modest houses with the help of the homeowners and volunteer labor. Donations of money and materials enable these Habitat houses to be sold to eligible families at no profit and to be financed with affordable loans. Through HFHI, many college students participate in building houses from the ground up, spending their spring and summer breaks building a Humanity home for members in their own community or abroad. Working on a construction site as a volunteer provides the best lessons in learning how homes get built. For further information, visit the HFHI website at www.habitat.org.

Volunteers construct a low-income home during a Habitat for Humanity project in Los Angeles.

FIGURE 1.21 The first and second stories of this home are built from CMUs that meet code requirements for hurricane-prone areas.

MASONRY WALL SYSTEMS

In modern masonry construction, exterior walls are made from stacked and mortared CMUs (Figure 1.21). The inner cavities of the concrete blocks are fitted with rebar and filled with concrete to provide additional strength and stability as the height of the wall increases (Figure 1.22). Masonry construction is typically seen in areas where there are frequent hurricanes. In fact, strict hurricane codes are

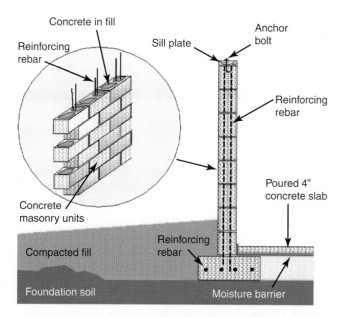

FIGURE 1.22 This section drawing shows the coursings of a masonry wall. Concrete foundation footings are reinforced with rebar, and the CMU walls are stabilized with rebar concrete infill.

FIGURE 1.23 CMUs are set with staggered joints (called a *running bond*) and cement mortar.

enforced in coastal areas of Florida that require the first floor, and often the second floor, to have masonry construction. Masonry wall systems like those made from CMUs, bricks, or structural clay tile are finished with a coating of durable *stucco* made from a mixture of *plaster* and cement that is weather resistive and can be painted (Figure 1.23). Masonry interior walls are finished with *drywall* or an application of plaster, which is then ready for painting or wall covering (Figure 1.24).

FIGURE 1.24 This Mediterranean–style home in coastal Florida is built from CMUs finished with painted stucco.

Roof Systems

The roof system is the most important component for weatherproofing a building structure. Regardless of style—whether multi- or single gabled, Dutch gambrel, shed, or flat—the design and construction of a roof must hold up against regional weather conditions, such as strong winds from tornados and hurricanes or the weight of heavy snow (Figure 1.25). The early American colonists of New England built their homes with a steeply pitched roof that allowed snow to slide off more easily during thaw and refreezing cycles. In fact, colonial-style homes are recognized by their prominent, steeply gabled roof. Frank Lloyd Wright's prairie-style homes, or a contemporary 1950s ranch house, would look odd without its low, flat roof. Traditional-style homes are more popular today, and the more interesting multigabled roofs that feature peaks and valleys over jut-outs of a garage, porch, or family room, for example, pose challenges for structural framing (Figure 1.26).

There are two methods of roof framing that are most common in residential construction: (1) roof trusses that are *prefabricated* offsite, then hoisted and anchored in place to the outer supporting walls, and (2) site-built roof rafters (Figure 1.27). Prefabricated roof trusses are carefully calculated for pitch and slope, and are sized according to the space they will cover, which cuts down on timber waste and ensures a perfect fit.

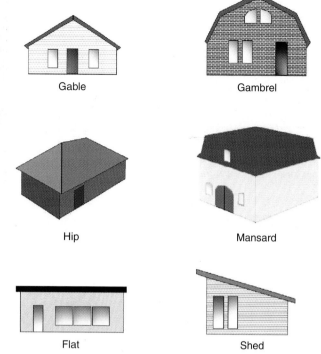

Gable

Gambrel

Hip

Mansard

Flat

Shed

FIGURE 1.25 This drawing shows a sample of the more popular roof designs featured on homes throughout the United States.

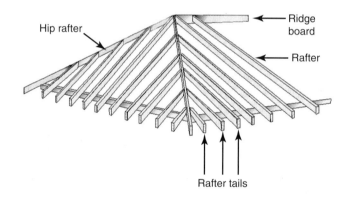

Hip rafter

Ridge board

Rafter

Rafter tails

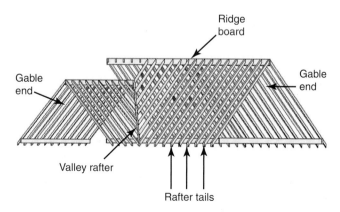

Ridge board

Gable end

Gable end

Valley rafter

Rafter tails

FIGURE 1.26 These diagrams show the framework and trussing systems for a hipped roof (top) and one with gables (bottom).

CAUTION

Hurricane Codes and Roofing

Building codes in hurricane-prone areas of seacoast communities require that roof trusses and rafters have additional *hurricane straps* installed to provide protection from strong winds. Hurricane straps made from galvanized steel are attached to one end of the roof rafters and are then attached to the exterior supporting walls. These straps are designed to hold the roof tight to the structural framing, minimizing the chance for a roof to be blown off during a hurricane. A similar strapping device is used to anchor the exterior walls to the foundation. For further information, consult the Federal Alliance for Safe Homes at www.flash.org.

A home in Punta Gorda, Florida, hit by Hurricane Charley in 2004. The home's exterior cladding has been blown away and the roof has collapsed. Hurricane straps are visible just above the large picture window, still attached to the exterior wall.

FIGURE 1.27 A crane hoists a prefabricated roof truss onto a timber frame structure.

FIGURE 1.29 At a new subdivision of houses in New Mexico, a construction worker attaches asphalt shingles to the roof. A layer of tar paper is laid over OSB cladding before the shingles are nailed in place.

FIGURE 1.28 Roof trusses form the shape for a gabled roof.

Trusses and rafters form the shape of the roofline and are the foundation for the finishing material (Figure 1.28). OSB or plywood sheathing is attached to the exterior roof, setting a strong substrate foundation for the final finishing material. A drip edge and *flashing* are applied along the perimeter of the roof, and a weatherproofing *underlayment* made of tar paper or polyethylene fiber is placed over the substrate material. The roof is then finished with roofing shingles, tiles, or shakes—materials chosen for their aesthetic appeal and their performance over the years.

Asphalt shingles are the most common material used in the roofing industry. They offer resistance to strong winds, are easy to repair, and are economical to purchase and install (Figure 1.29). Asphalt shingles are available in a variety of colors, and manufacturers guarantee them for 20 or 30 years. Wood shake roofing shingles made from cedar, redwood, or southern pine offer a more expensive alternative to asphalt (Figure 1.30). Although the natural beauty of wood shakes has great aesthetic appeal, this material is not *fire resistive* and does not meet municipal codes in certain regions like southern California, where wildfires are common. Fire protective coatings can be applied to wood shingles, which afford some protection, although *engineered wood* products that look like the real thing are readily available.

Natural slate is a roofing material that lasts for 75 to 100 years and was a popular choice as a roofing material during the late 19th century. When constructing the roof trusses for a slate roof, its framing must be reinforced to support the heavy load. Slate is available in naturally variegated colors and is the perfect choice for a vintage-style home. Natural slate tiles cost more than wood shakes to purchase and install, but today's manufacturers are offering less expensive alternatives made from recycled products—like old rubber tires—that look like real slate (Figure 1.31). Similarly, roof trussing systems must be planned beforehand when using concrete

FIGURE 1.30 Cedar shingles are nailed in place over a tarpaper underlayment on this new house in Maryland. The worker begins at the bottom of the roof and works up to the ridgeline, overlapping each row of shingles to ensure a weather-tight fit.

FIGURE 1.31 The irregularity of color and texture in these faux slate shingles gives the appearance of the real thing.

FIGURE 1.32 A worker nails concrete shingles onto strips of wood lathing that are firmly attached over tar paper.

shingles or those made from ceramic (Figures 1.32 and 1.33). Manufacturers of concrete and ceramic roofing tiles offer virtually limitless combinations of texture and color for creating interesting effects.

It is important to note that roofing materials like slate, concrete, and ceramic are inherently *noncombustible* and resist the spread of fire. Although these types of roofing materials cost more initially, they last longer than asphalt or wood shakes. Furthermore, roofing shingles made from fiber cement are actually better for the environment, because they are made from natural materials and recycled *postproduction waste*.

The popularity of metal roofs has come a long way since colonial times, when corrugated metal roofs covered barns, sheds, and farmhouses. An increasingly admired alternative to traditional roofing materials, metal roofs are available in enamel-painted finishes, and are long lasting with guarantees to perform for up to 100 years (Figure 1.34). Although metal roofs are extremely durable, the initial cost to fabricate and install them makes them more expensive than traditional asphalt shingles.

FIGURE 1.33 Ceramic roofing tiles, often called *barrel tiles*, cover the roof of this Florida home.

LEED the Way

Leadership in Energy and Environmental Design

Established in 1993 as a nonprofit organization promoting sustainable building practices, the U.S. Green Building Council (USGBC) instituted the Leadership in Energy and Environmental Design (LEED) Green Building Rating System to certify construction projects that meet strict criteria for protecting the environment. According to the organization's website, LEED rates residential home construction projects based on five areas of criteria: sustainable site development, implementation of water-saving systems, increased energy efficiency, integration of sustainable or recyclable materials, and indoor environmental quality. For further information, visit the USGBC website at www.usgbc.org to download "LEED for Homes Rating System" (www.usgbc.org/ShowFile. aspx?DocumentID=3638).

FIGURE 1.34 A metal roof covers this multigabled roof. The roof sections are prefabricated, cut to the required length, powder coat painted, and then installed on site.

Roofing

 Manufacturers of roofing materials are looking for ways to "go green" these days in an effort to save the environment and to increase profits by turning readily available resources like recyclable materials into roofing products. Reengineered roofing shakes are made from 90% to 100% recycled materials and include rubber, vinyl, plastic, and other postindustrial products. Shaped to look like cedar shakes or natural slate, these products are economical and endure the test of time, with most manufacturers offering a minimum 50-year warranty.

Another milestone in ecofriendly roofing includes solar panels designed to look like slate roofing tiles or asphalt shingles. These innovative products are currently finding a market for homeowners who want lower electricity bills. The energy, produced by *photovoltaic* cells, is channeled via connector wires to an inverter box near the home's electrical panel box; the energy is then converted to provide electrical power for the home.

Roofers install slate-style solar tiles on this new home.

Floor Systems

The floor of a home is more than just carpeting, tile, or hardwood; it begins with the foundation. For slab-on-grade foundations, the concrete slab is the subfloor inside the house. Because slab-on-grade foundations are poured in place over a moisture barrier, finish materials are protected against damage by groundwater seepage and moisture. Pier-and-beam, crawl space, basement, and stem wall foundation systems use *floor joists* that are usually two-by-sixes or two-by-eights, depending on the load-bearing requirements and span lengths. These joists are attached to the foundation walls. Wood floor joists provide the structural support for attaching a subfloor of OSB or plywood that will then be covered with the selected finish material such as carpet, tile, or wood (Figure 1.35).

In addition, floor joists are spaced every 12, 16, or 24 inches on center, with the spacing determined by the structural load the floor system will carry and the weight of the interior finish materials (Figure 1.36). For example, stone and ceramic tiles are heavier than resilient vinyl flooring or carpeting, so the floor joists must be calculated and configured in span, spacing, and dimension to support the weight of these heavier materials. This floor system is repeated for the upper levels of the house, and the floor joists on the second story are actually the ceiling joists for the floor below. Ceiling materials and finishing processes are discussed in more detail in Chapter 3.

FIGURE 1.35 OSB covers laminated veneer lumber floor joists, which become the structural support system for the finished flooring material.

FIGURE 1.36 Floor joists of laminated veneer lumber are engineered from wood composites to provide extra strength and durability than traditional lumber. These two-by-eight floor joists are spaced every 16 inches on center.

Doors and Windows

No house is complete without doors and windows. In choosing windows and doors for a new home, considerations include style, performance, and security, and are decided upon during the initial stages of home design. A door and window schedule is prepared by the architect that lists the manufacturer, model number, type, and *rough opening dimensions*. This schedule becomes part of the set of construction documents mentioned at the beginning of this chapter, and is used by the contractor to ensure all framed-out openings will accommodate the specified doors and windows. The exterior framing is guided by the size and locations of the doors and windows as listed on the schedule.

The door schedule is keyed to the plan by using a sequential listing of numbers for easy cross-referencing for locating the right type of door to the correct opening. Similarly, windows are coded with letters (Figure 1.37). Doors and windows must be selected according to their energy efficiency ratings to meet local energy codes.

Door schedules, in addition to listing the description and dimensions of the door and hardware needs, include details of the handing (or, which side of the door has hinges) and the direction of the door swing. For example, standing outside and

CAUTION

Hurricane Codes and Doors

In hurricane-prone areas, residential doors must swing away from, rather than in to, the home. The door is prevented from opening to the inside by its casing and hinges, which gives the door strength to withstand the pressure of strong winds against its surface.

DOOR SCHEDULE							
DOOR #	DOOR OPENING			MATERIAL	FINISH	HARDWARE	NOTES
	HEIGHT	WIDTH	THICKNESS				
1	6'-8"	3'-0"	1-3/4"	WD/GLASS	STAIN/PAINT	LOCKSET	1. SOLID CORE 2. STAIN ON EXTERIOR, PAINT ON INTERIOR SIDE AS SPECIFIED
2	6'-8"	2'-6"	1-3/4"	WD	PAINT AS SPECIFIED	PASS THROUGH	1. PRE-PRIMED, HOLLOW CORE
3	6'-8"	6'-0"	1-3/4"	GLASS SET IN WD FRAME	PRE-PAINTED	LOCKSET	1. ANDERSEN SLIDING 400 SERIES
4	6'-8"	3'-0"	1-3/4"	MT'L	PRE-PAINTED	LOCKSET	
5	6'-8"	2'-8"	1-3/4"	WD	PAINT AS SPECIFIED	LOCKSET	1. PRE-PRIMED, HOLLOW CORE
6	6'-8"	3'-0"	1-3/4"	WD	PAINT AS SPECIFIED	PASS THROUGH	1. PRE-PRIMED, HOLLOW CORE

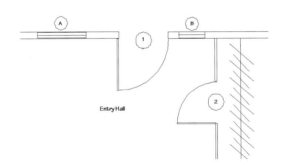

Entry Hall

WINDOW SCHEDULE							
WINDOW SYMBOL	ROUGH OPENING			TYPE	FINISH	MFG & MODEL #	NOTES
A	36-3/4"	60-3/4"	3/4"	WD/GLASS DOUBLE HUNG	STAIN/PAINT	ANDERSEN 400 SERIES	1. STAIN ON EXTERIOR, PAINT ON INTERIOR SIDE AS SPECIFIED
B	64-3/4"	12-3/4"	3/4"	WD/GLASS STATIONARY	STAIN/PAINT	ANDERSEN 400 SERIES	1. STAIN ON EXTERIOR, PAINT ON INTERIOR SIDE AS SPECIFIED
C	48-3/4"	60-3/4"	3/4"	WD/GLASS SET SLIDING	STAIN/PAINT	ANDERSEN 400 SERIES	1. STAIN ON EXTERIOR, PAINT ON INTERIOR SIDE AS SPECIFIED

FIGURE 1.37 A combination of symbols, numbers, and letters is used on a floor plan to convey the type of doors and windows to be incorporated, and these codes are cross-referenced to a schedule.

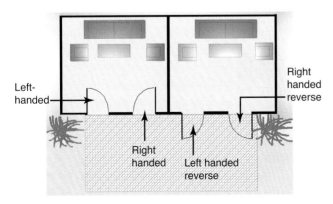

FIGURE 1.38 This illustration shows how *door handing* is assigned in preparation of a door schedule.

facing the front door, if the hinges are on the left and the door swings toward the interior, it is a left-handed door. However, if this door is hinged on the left but opens to the outside, it is noted as a left-handed reverse door (Figure 1.38).

Studs for door and window locations are framed out with double or sometimes triple studs on the sides surrounding the openings and include a framed-out *header* above for added structural support (Figure 1.39). Structural framing for door installation is determined by the type of door and whether it is an exterior or an interior door. Framing for interior doors does not require the double and sometimes triple stud reinforcements required for exterior doors; exterior doors must be framed to resist break-ins. For masonry construction, CMUs are sized and stacked, leaving openings for windows

FIGURE 1.39 Openings for doors and windows are formed by tightly grouping wood studs to reinforce openings in load-bearing walls. The framing for an interior door features a header stud at the top for stabilizing the framing.

FIGURE 1.40 CMUs are sized and stacked according to locations for doors and windows. The door and window assemblies are attached to a wooden frame surrounding the openings.

and doors that are then lined with wood framing for fastening the door or window *assembly* (Figure 1.40). Doors for new construction are available as prepackaged door assemblies that are shipped to the site complete with door, *jambs*, and hardware. Exterior doors are then trimmed out on both sides with the appropriate wall finish materials (Figure 1.41).

Like doors, windows need special framing to perform over the many openings and closings they will endure with daily use (Figure 1.42). Prepackaged window sets include the side jambs, sill, and framing that keeps the glass in place.

DOOR TYPES AND STYLES

Architects and interior designers choose doors according to their performance in the home. Exterior doors must be able to keep the weather out—along with cold or warm air—and protect the homeowner against break-ins. The most common types of exterior doors are those that swing and those that slide (Figure 1.43). Interior doors provide *acoustical* control and privacy, and include specialty types like pocket doors that slide into the wall, bifold doors that are hinged in sections like the folds of an accordion, and surface sliding types

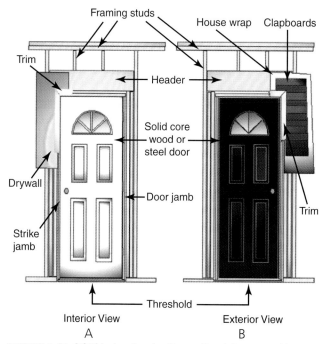

FIGURE 1.41 (A) This drawing details a wall and door assembly as seen from the interior space. (B) This drawing details these same elements from the exterior.

FIGURE 1.42 A rough opening for a window is trimmed out with one-by-twos.

FIGURE 1.43 Sliding doors with large panes of fixed glass offer an unobstructed view to the outside and eliminate the space required for swinging doors.

(Figure 1.44). Specialty doors offer solutions when space is limited, because swinging doors need space for the door to open and close without obstruction. Bifold doors reduce the amount of space needed to open and close by the number of folds in the door itself, and pocket doors disappear into wall cavities, keeping openings clear of obstructions.

Exterior and interior doors come in a range of materials, sizes, and styles, fitting any aesthetic (Figure 1.45). Glazed doors feature large panels of *safety glass* like those used for patio doors, sash doors have small sections of glass, and more traditional doors feature raised panel designs. Doors specified in residential construction vary between solid-core wood doors, hollow-core wood doors, vinyl doors, or metal-clad doors. Each of these types has advantages and disadvantages. For example, a hollow-core door costs less than a solid-core

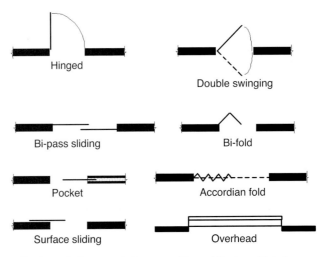

Hinged

Double swinging

Bi-pass sliding

Bi-fold

Pocket

Accordian fold

Surface sliding

Overhead

FIGURE 1.44 These symbols are used by architects and interior designers on floor plans to designate the type of door that will be specified; these are common door types used in residential projects.

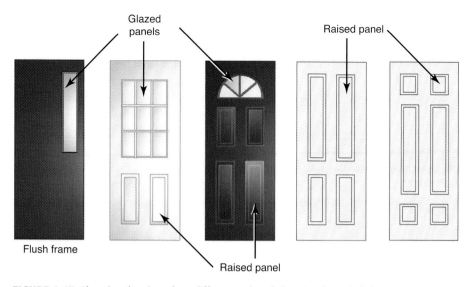

FIGURE 1.45 *Elevation drawings* show different styles of glazed and paneled doors.

door, but is not sturdy enough to be used on the exterior. Exterior doors made with a solid-core material resist the impact of a break-in and provide important thermal control. The performance of a vinyl door over a wood door is better in areas where high humidity might otherwise expand the wood frame, causing the door to stick.

According to the U.S. Department of Energy, exterior doors meeting the requirements for Energy Star certification must provide maximum insulation, must be tight fitting to reduce air leaks, and must be made from fiberglass, solid-core wood, or steel cladding over a polyurethane foam core. Exterior doors with *glazing* must have either double or triple layers of insulating glass panes to be Energy Star certified.

Door hardware includes the hinges on which the door is hung, door knobs or levers for opening and closing, and locking devices when needed for privacy or security. Exterior doors and interior doors—specifically those to bathrooms—are specified with lock sets. Interior doors that do not require privacy (like closets) are fitted with pass-through sets. Specifications for all hardware are listed on the door schedule included in the set of construction documents.

WINDOW TYPES AND STYLES

Providing views to the outside, windows are one of the most prominent features of a new home and are chosen to maintain the look of the designed style. *Muntin* windows fit right into a colonial-style home, but would look out of place on a 1950s split-level ranch. Sash windows are those with glass panels that open from the bottom (single hung) or from both the bottom and the top (double hung). Opening both the top and bottom sash in a double-hung window allows for better air circulation; an opened lower sash brings in fresh air whereas the upper sash draws out warmer air that has risen toward the ceiling (Figure 1.46). Casement windows are hinged on their side and crank open to capture breezes and direct them inside (Figure 1.47). Casement windows, like swinging doors, need room without any obstruction for opening and closing. Awning windows also crank open, but are hinged at the top. These windows are commonly used for the upper windows in basements or in sloped ceilings, and their hinged top is designed to keep out the rain (Figure 1.48). Sliding windows open by sliding one panel over another, whereas fixed windows do not open at all.

After windows are installed in new construction, they are finished with decorative trim on both the exterior and interior. Like doors, wood, vinyl, and metal windows have advantages and disadvantages. Consult the manufacturer's warranty to

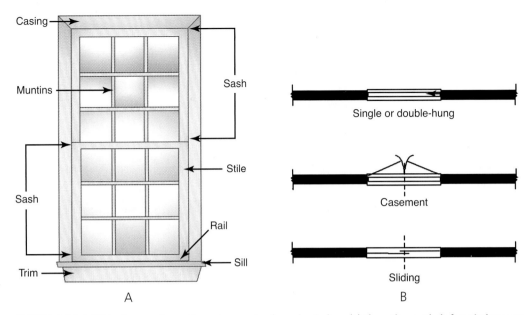

A

B

FIGURE 1.46 (A) This diagram shows the various parts of a sash window. (B) Floor plan symbols for windows are shown in this drawing.

FIGURE 1.47 Casement windows are hinged on the side and crank open to the outside to direct breezes into the interior.

FIGURE 1.48 Awning windows become skylights in the roof of this residence.

LEARN More

Reengineered Houses for the 21st Century

Many of the construction materials used in today's single-family homes are prefabricated and engineered for the betterment of the environment. Using prefabricated and engineered materials made from recycled or surplus postproduction materials cuts down on construction waste and saves our natural resources. The United States Environmental Protection Agency reported that in 2008, 33% of our nation's solid waste came from building construction (www.epa.gov/wastes/partnerships/wastewise/pubs/progrpts/pdfs/report14.pdf). Some of the resource-saving products specified for residential construction include OSB sheathing, roofing shingles, siding materials, and window frames. OSB sheathing saves trees, because it is made from postproduction wood waste. In addition, roofing shingles made from recycled rubber tires are formed to look like wood shakes or slate tiles, and siding materials like clapboards made from wood composites, or faux brick made from fiber cement all help save natural resources. Energy-efficient windows feature recyclable vinyl frames or frames made from wood composite materials. For further information on specifying resource-saving products, visit the green section of the National Association of Home Builders

website at www.nahbgreen.org. For more information on engineered wood products, visit www.apawood.org: APA—The Engineered Wood Association website.

Designed in a transitional style with a hipped roof, large expanses of glass windows, and an arched entryway, this new home was built with the most up-to-date, engineered materials. The APA-rated 303 siding panels on the home's exterior provide excellent natural thermal insulation properties and look like natural wood.

determine which window material is best suited for the job, for example, windows with wood frames do not perform well in areas where humidity is high.

Within the past two decades, cities and towns across the country have established building codes that require windows in new construction projects to meet minimum guidelines for efficiency. Since then, a variety of energy-efficient windows earning the Energy Star seal have entered the market. Varieties range from those designed with an air pocket sealed between two layers of glass to reduce air infiltration and provide greater insulation, to those with a single layer of glass coated with thermal insulating *E-coatings*. For more information, visit the Energy Star website for helpful tips (www.energystar.gov).

resources

Informative Websites

APA—The Engineered Wood Association: www.apawood.org

Energy Star: www.energystar.gov

Federal Alliance for Safe Homes: www.flash.org

Habitat for Humanity International: www.habitat.org

International Code Council: www.iccsafe.org

LEED for Homes Rating System:
www.usgbc.org/ShowFile.aspx?DocumentID=3638

National Association of Home Builders, Green section:
www.nahbgreen.org

Northeast Sustainable Energy Association: www.nesea.org

Portland Cement Association: www.cement.org

U.S. Department of Energy—Codes: www.energycodes.gov

U.S. Green Building Council: www.usgbc.org

CHAPTER 2

Systems for Residential Construction

Electrical, Heating and Air-Conditioning, and Plumbing Systems

Integral to the basic functioning of life within the home are electricity, heating and air-conditioning, plumbing, and communications wiring for telephones, Internet, and security systems. These elements are often planned before the interior designer gets involved in a project; yet, an interior designer needs to review the locations of these systems with the building contractor and architect to ensure they are integrated into the aesthetic scheme of the home. Locations for housing heating and air-conditioning equipment are limited to closets, attics, or basements, but the placement of supply and return air grills within the home can be negotiated if these elements interfere with a large built-in entertainment center, for example.

Furthermore, a thorough review of the lighting and electrical plan by an interior designer ensures options like ceiling fan locations and floor outlets for furniture groupings are identified before construction begins. Special plumbing lines servicing

This modern adobe home features solar collecting panels on the roof and a wind turbine. These alternative energy resources supply the home with electricity.

a pot filler over the kitchen stove require piping that must be planned for in advance, before walls and ceilings are enclosed by *gypsum wallboard* or other finish materials. These important systems affect the interior space, and designers must have a basic understanding of these features when working on a construction project.

Electrical Systems

Thomas Edison (1847–1931) demonstrated his invention for an incandescent light bulb as early as 1879, but having electricity wired in a home during the construction process did not occur until the turn of the 20th century. In fact, some of the first homes designed with fully integrated electrical systems were the modest bungalow-style houses built across America during the early 1900s. In those days, the wiring system was primitive compared with today's standards and is considered a fire hazard. A network of wires ran through ceramic tubes or around ceramic knobs attached to interior framing studs. These ceramic knob-and-tube insulators kept the wiring from making contact with the wood studs to prevent fire. Knob-and-tube wiring is illegal in many states and needs to be upgraded during home renovations. In addition to the old knob-and-tube system, during the mid 20th century, aluminum wiring was used in new home construction. Although this type of wiring is still functional, the risk of fire makes it worth replacing with more modern copper wiring.

Electricity is brought into the home from the main utility supply, either from visible power poles and lines or from components buried underground. A main service *conductor* attached to a meter measures the usage of electricity in *kilowatt* hours. This meter connects to a main panel box, often called a *circuit breaker,* which is located inside the home (Figure 2.1). In residential construction, the panel box is equipped to carry between 100 amps and 200 amps (the rate at which electricity is delivered). Currently, in the age of multiple computers and printers, fax machines, surround sound equipment, and big-screen TVs, newer homes are usually equipped with 200-amp panel boxes.

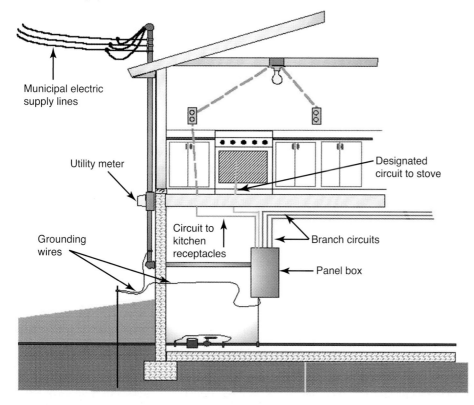

Municipal electric
supply lines

Utility meter

Grounding
wires

Circuit to
kitchen
receptacles

Designated
circuit to stove

Branch circuits

Panel box

FIGURE 2.1 This diagram shows how electricity is brought into the home from the main supply and is connected to a circuit breaker box in a basement, where electricity is then distributed through branch circuits to specific areas.

LEARN More

Alternative Energy

The burning of fossil fuels for the past several centuries has damaged the environment and has reached critical stages in upsetting the ecological balance of the planet. Alternative energy sources like wind power are playing a major role in providing a safe source for generating power. In 2010, the United States produced more than 35,600 megawatts of productive wind generation—enough to power the equivalent of 9.7 million American households annually (American Wind Energy Association, 2010). According to the report "20% Wind Energy by 2030: Increasing Wind Energy's Contribution to U.S. Electricity Supply," which was released by the U.S. Department of Energy in 2008, wind power is capable of becoming a major contributor to America's electricity supply during the next three decades. Continued government support through tax breaks and other incentives are helping to accelerate wind energy development throughout the nation. An increasingly competitive source of energy, wind turbines or windmills could provide at least 20% of the nation's electricity without consuming any

Wind turbines generate electricity channeled to the city grid.

natural resources, and without emitting any pollutants or greenhouse gases. Learn more by visiting the website of the American Wind Energy Association at www.awea.org.

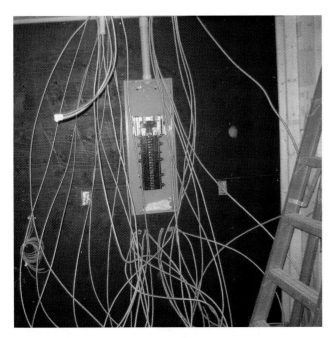

FIGURE 2.2 A panel box in this new home is ready for the electrician to connect wiring to individual circuits that will be distributed throughout the home. Each circuit will carry from 15 to 20 amps.

CIRCUITS

After the main power is supplied to the panel box, the electricity is divided into *branch circuits*. Each branch supplies electricity to a series of *receptacles* (often called *outlets*) and switches for one area or zone of the house (Figure 2.2). For example, receptacle outlets in a kitchen or home office are usually separated from other zones and are put onto designated circuits. Separating these high-electrical- use areas onto separate designated circuits prevents the risk of overloading the electrical system. Furthermore, placing specialized equipment or appliances on designated circuits reduces the chance of a *short circuit* caused by having several appliances or equipment in operation at the same time. Circuits are controlled by breakers inside the panel box, which cut off the flow of electricity in case of an overload. Essentially, the electricity to that circuit is shut off; the breaker automatically switches or "trips" to the "off" position, minimizing damage to appliances and equipment or the wiring system.

The consistency of the *electric current* running through the circuit, or loop, is measured in *amperes*. Individual circuits are equipped to handle between 15 amps and 20 amps each (Figure 2.3). A reduction in the amperage of a circuit as a result of an overload at the municipal supply is commonly called a *brownout;* *blackouts* are the total loss of electrical power. Using surge protectors on electronic equipment reduces the chance of damage in case of brown- or blackouts. The flow of electricity through the circuits is controlled by switches—either those mounted on a wall, like the ones that turn lights on and off, or those that are built directly into an appliance. A switch in the "on" position allows the flow of electricity to feed power to the light, appliance, or equipment that is "plugged in," so to speak, to the circuit (Figure 2.4).

Branch circuit wiring is channeled from the panel box through drilled holes in the stud walls (Figure 2.5). *Junction boxes* for switches and receptacles are mounted

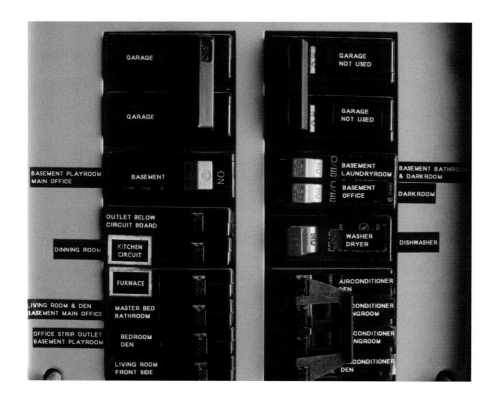

FIGURE 2.3 Individual circuit breakers are labeled for each room or area of the home. The breaker will automatically switch to the "off" position if a circuit becomes overloaded.

with screws to the side of the studs or onto ceiling or floor joists (Figure 2.6). The National Electrical Code (NEC) regulates how many receptacles are required for specific rooms based on length of wall space, and also sets the mounting height from the floor for these outlets. Specifically, the NEC states that receptacle outlets must be no farther than 10 feet apart in rooms other than kitchens. Kitchen spaces are required, by code, to provide a receptacle every 4 feet, and although wall outlets are mounted 15 inches *above finished floor (AFF),* those above kitchen counters are mounted 42 inches AFF, which is high enough to clear 36-inch AFF countertops.

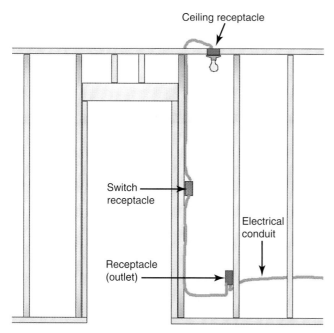

FIGURE 2.4 Wiring for an overhead light and controlling switch inside a room is shown. Electrical wiring runs from the circuit breaker to designated outlets and light fixtures. The wall switch interrupts the flow of electricity, controlling the "off and on" of the lamp.

FIGURE 2.5 The *rough-in* for a wall outlet shows how the electrical wiring is fed from the panel box in the basement up through holes drilled into the sill plate and stud walls, and pulled through to the location of the wall receptacle.

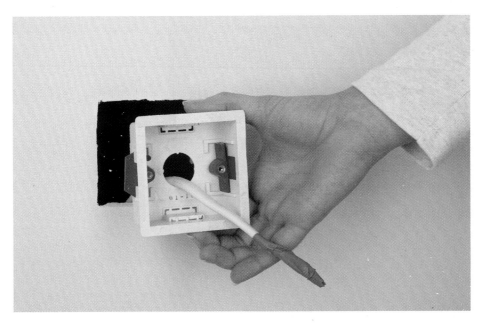

FIGURE 2.6 The finished drywall is cut around a plastic junction box for Internet cabling and a data port.

FIGURE 2.7 An electrician installs the housing for a ceiling light.

Ceiling outlets channel electricity to run fans and operate lights that are "hardwired" into the circuitry, whereas wall or floor receptacles accommodate appliances or equipment that are plugged into the outlets (Figure 2.7). In the United States, the *voltage* brought into most residential receptacle outlets is 120 volts. Special equipment or appliances such as ovens or ranges and electric clothes dryers run on designated circuits supplying 220 volts of electricity. These appliances are usually hardwired to the circuit. By code, receptacles installed near wet walls, like sinks, must be equipped with interrupters to stop the flow of electricity in case water accidentally makes contact with the wiring. This precaution prevents electrical current from coming into contact with water. A *ground fault circuit interrupter* (or GFCI) is required near bathroom or kitchen sinks per the NEC. In addition, outdoor receptacles must also include ground fault interrupters and have a protective plastic covering.

CAUTION

Electrical Codes

Municipal codes regulating residential construction enforce compliance with the guidelines established by the NEC. An electronic copy of this code is available through the National Fire Protection Agency's website (www.nfpa.org) and provides valuable information, such as the amperage limits for branch circuiting, the required distance between outlets in kitchens, where light switches should be placed in a stairwell, and which outlets must have GFCI protection. For instance, these codes require that staircases must have light switches at the bottom and top of the staircase so that the stairwell is safely lit from either direction. Also, rooms with two or more doorways require a switch at each point of entry into the room. Furthermore, the NEC provides installation requirements for electrical equipment, including breaker boxes, receptacle outlets, and switching, and for hardwiring heating and air-conditioning equipment.

FIGURE 2.8 Fiber optic cable, the latest technological advancement in communications wiring, transfers information through the fibers with light.

LOW-VOLTAGE WIRING SYSTEMS

The communications industry is rapidly changing and planning for telephones, *asymmetric digital subscriber lines (ADSL),* intercoms, surround sound systems for televisions, and home security systems—all of which require more specialized planning. Well-designed homes are now fitted with data ports in every room in the house, anticipating what the future will bring to communication needs. Data ports are connected to a centralized control system where phone lines, ADSL communication, and high-definition or satellite television cables are brought into the home from the main utility lines of the service provider. Newer technology using *fiber optic cable*—thin strands of glass transmitting data at high rates of speed through beams of light—is beginning to replace copper coaxial wiring in telephone and broadband communications (Figure 2.8).

Low-voltage wiring systems for home security systems require the services of trained professionals who are involved with the project early, usually when the lighting and electrical plans are prepared. All electrical wiring must be installed before the walls and ceilings are enclosed with gypsum wallboard or other finish materials (Figure 2.9). Advances in the home security industry enable homeowners to monitor home activities from a remote location. Current technology allows the homeowner to turn lights on and off, regulate the air-conditioning or heating, open or close drapes, and turn on televisions or stereos from a laptop computer.

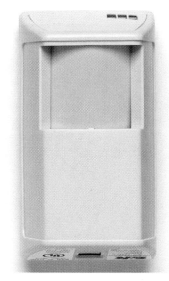

FIGURE 2.9 Motion detectors are hardwired into the electrical supply and mounted in key areas throughout the home.

CAUTION

Smoke Detectors

Smoke detectors emit a high-pitched alarm to alert people to the possibility of fire.

National Fire Protection Agency (NFPA) codes require new homes to have smoke detectors hardwired to the main electrical service and to be supplied with a backup battery in case of blackouts. The code requires that at least one smoke detector be installed on each level of the home, and smoke alarms are required for all sleeping areas within the home. Alarms placed in adjoining hallways to bedrooms must be audible through closed doors. Visit the NFPA website for further information (www.nfpa.org).

LIGHTING AND ELECTRICAL PLANS

A typical lighting and electrical plan prepared by an architect or interior designer diagrams the locations of all ceiling, wall, and floor receptacles; special outlets required for high-voltage appliances; communication ports needed for telephone, Internet, and cable; and wall switches controlling any or all of these receptacles (Figure 2.10). It is important for the interior designer to consider the locations of furniture and equipment when designing a lighting and electrical plan. Outlets may need to be located in the floor to accommodate a large room where furniture arrangements are grouped away from walls. In addition to data ports, switches, receptacles, and

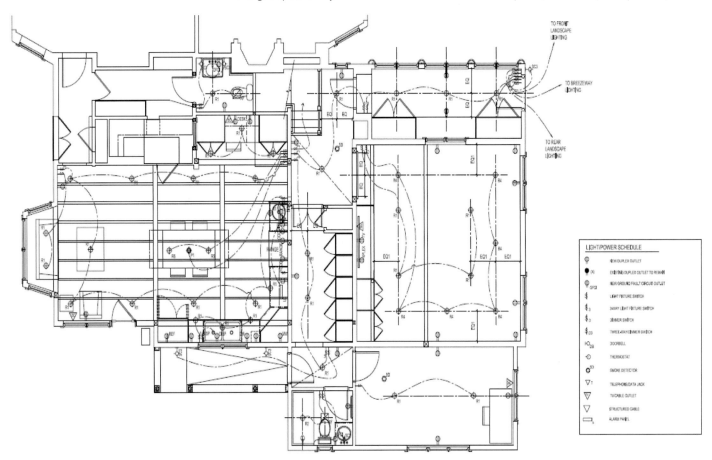

FIGURE 2.10 A lighting and electrical plan shows the locations of all ceiling lights, switches, and wall outlets, including those for communication and media. The drawing is accompanied by a schedule of symbols to ensure accurate identification by the installer.

smoke detectors, the designer will check the locations of thermostats in rooms that are wired to heating and air-conditioning systems to ensure that these locations coordinate with built-ins like bookcases or entertainment units. Altering any locations for these devices must be discussed and approved by the electrician and *general contractor* to ensure NEC compliance.

Heating, Ventilation, and Air-Conditioning Systems

Heating and air-conditioning systems are an essential part of new-home construction. HVAC systems must maintain thermal comfort for the occupants, be energy efficient (in some regions this is required by code), and be economical to operate. Regional weather patterns determine which system is best suited for providing efficient thermal control. In cold climates, gas or propane fuels feed boiler units for water or steam radiant heating systems; in mild climates, forced-air heating systems run off electricity. Regardless of whether the system is forced air or some form of radiant heating, equipment is sized according to the square footage of the home and is measured in BTUs. (*British thermal units* measure thermal output.)

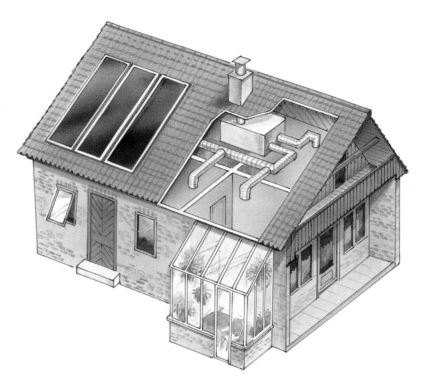

FIGURE 2.11 This cutaway drawing shows how roof-mounted solar panels supply power to an attic-mounted HVAC system. The system delivers hot or cold air to rooms inside the home via a network of ducts.

HVAC systems are separated into multiple zones designed to provide the best thermal comfort throughout the house. Individual thermostats maintain preset temperatures in each zone. Areas like kitchens, where cooking and baking increase room temperature, should be considered a separate zone. Each type of HVAC system has its advantages and disadvantages, and deciding which system is best for a new home depends on operational costs and the effectiveness of the system relative to geographic location.

FORCED-AIR SYSTEMS

Air handlers with a refrigeration compressor, and heating mechanisms like a furnace, deliver heating or cooling through a forced-air system. Air-handling refrigeration units are usually located outside, so the noise is kept away from the inside of the home (although it is not uncommon for these units to be installed in attics or ventilated basements). Forced-air systems bring warm or cool air produced from a heat pump or a refrigeration compressor into the home through a planned system of ductwork (Figure 2.11). Ductwork occupies a sizable amount of space and is hidden in wall cavities, ceilings, and floors (Figure 2.12).

Forced central air systems rely on carefully placed supply-and-return grills to maintain airflow and healthy ventilation. Fresh air is pumped into the air handler distributed through the house via supply ducts. Return grills located inside the home suck out stale air, usually through vents in the attic. In most cases, return grills are placed low to the floor whereas supply vents are located near ceilings (Figure 2.13). An advantage of installing forced-air systems for heating and cooling is that they control air movement and are equipped to maintain relative humidity. *Indoor air quality* is an important consideration when using forced-air systems. The filters installed with these systems reduce the amount of dust, odors, and pollens circulated through the interior, but only if properly maintained via regular cleaning or replacement of filters.

FIGURE 2.12 The HVAC ductwork in this new construction project will be hidden behind the finished ceiling. Supply vents deliver air from the compressor unit to key positions throughout the interior.

The federal government has issued energy standards for manufacturers of home appliances and building products like windows and doors. The Energy Star seal of approval is given to those products that meet the standards prescribed by the government for minimizing greenhouse gases either through their production or construction methods. For a new home to receive the Energy Star seal from the federal government, the house must be constructed in a way that achieves at least 15% more energy efficiency than other homes built according to the 2004 International Residential Code. This rating reflects that a home was constructed with sufficient insulation, efficient heating and cooling equipment, and Energy Star–certified lighting and appliances. For more information, visit the Energy Star website at www.energystar.gov.

The Energy Star label appears on products designed to promote energy efficiency by reducing greenhouse gas emissions.

Purchasing products with the Energy Star emblem helps promote steps toward reducing negative effects associated with climate change.

Consumers are more familiar with the yellow Energy Guide label, which appears on home appliances ranging from refrigerators to washing machines and clothes dryers. The Energy Guide label allows consumers to compare the energy use of a particular product with others by offering detailed information regarding product performance. A scale shows how the labeled product compares with other models and provides estimated energy costs in U.S. dollars based on the yearly operation of the unit.

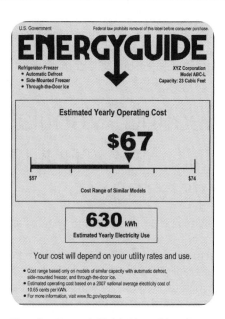

The yellow Energy Guide label for a dishwasher reports the efficiency rating of the unit along with comparisons with other models on the market. The Energy Guide also estimates the annual operating costs for the appliance.

WALL-MOUNTED SYSTEMS

Newer to the American market are all-in-one systems that require the least amount of space for running ducts and pipes (Figure 2.14). These systems have outdoor compressors that feed supply lines through small-diameter *conduits* in wall cavities that connect to a wall-mounted air handler. Although these newer systems are often criticized for their lack of aesthetic appeal, they prove to be economical to install. Moreover, all-in-one systems do not require a network of large ductwork hidden in ceilings and walls, making them good solutions for remodeling projects that had no prior system. More important, these units operate on up to 50% less energy than conventional units, and most models are equipped with self-cleaning filters.

RADIANT HEATING SYSTEMS

There are two types of radiant heating systems: those that rely on hot water or steam and those that are heated by an electrical current. Water-fed radiant heating systems have pipes that distribute hot water or steam from a *boiler* to radiators positioned throughout the house (Figure 2.15). Boiler-fed hot-water and steam

FIGURE 2.13 This diagram traces the connections for a condenser unit, furnace, and ductwork. Return grills located near the floor suck out stale air to maintain constant air movement.

Indoor Air Quality

Indoor air pollution is a growing concern among homeowners and businesses, because the harmful effects of breathing polluted air range from Legionnaire's disease to chronic allergy problems. A tightly sealed building contributes to poor indoor air quality by reducing (1) the amount of fresh air brought into the living environment and (2) the elimination of stale or polluted air from the interior. Unless precautions are taken, dust mites, mold, and microbial growth contribute to poor indoor air quality. The most important factor in combating poor indoor air quality is adequate ventilation—controlling air movement—including the filtration of air and the elimination of stale

An electron microscopic view shows what is commonly found in the air inside a typical family home: skin flakes, pet dander, and hair.

air through return grills. Properly ventilated homes help alleviate indoor pollution and, by controlling the interior temperature and relative humidity, greatly aid in reducing the growth of mold spores and dust mites. An environment of high humidity is a breeding ground for microbial growth. In some cases, toxic mold growth has led to the complete demolition of houses, because its damage cannot be undone. Good indoor air quality is achieved through proper and frequent cleaning or replacement of air filters and filtration systems.

Cleaning the HVAC ductwork on a regular basis removes potentially toxic airborne debris, improving indoor air quality.

heating systems initially cost more to install because of the cost of equipment (such as the radiator units) and copper plumbing pipes; however, they are more economical to operate when boilers are fired by fossil fuels or alternative energy sources. Radiant systems have been around for more than 100 years and still function in older homes. These systems need space for the boiler, a network of pipes, and radiators, which may obstruct otherwise usable space. Networks of pipes operating on a complete circuit from the boiler to radiator are divided into separate zones. When the temperature drops to a predetermined setting, a thermostat signals the

FIGURE 2.14 Wall-mounted air-conditioning and heating systems like this one are connected to a compressor located outside or on the roof. A small conduit runs from the compressor to the wall unit, which eliminates the need for internal ductwork. A remote control operates the unit, which is regulated by an internal thermostat.

FIGURE 2.15 The boiler heats water that runs through a circuit of pipes that connects to radiators in the house. The boiler unit also heats the water supplied to faucets, eliminating the need for a separate hot water heater.

FIGURE 2.16 A copper pipe connected to a boiler in the basement is brought up through a hole in the floor. When connected to the radiator, it will force hot water or steam from the boiler through the coils and through a circuit of pipes, returning the water to the boiler for reheating.

FIGURE 2.17 This baseboard heater works in the same way as standing radiators, yet it is less intrusive in the interior space. Baseboard heaters work off a boiler system or electricity.

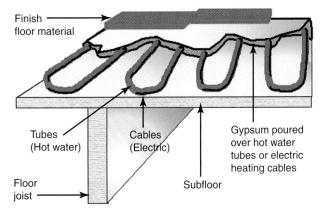

Finish floor material

Tubes (Hot water)

Cables (Electric)

Gypsum poured over hot water tubes or electric heating cables

Floor joist

Subfloor

FIGURE 2.18 Radiant flooring is installed on a pier-and-beam foundation system. The OSB or ¾-inch plywood subfloor supports tubing carrying the heat through a circuit. The heat can be electric or hot water, depending on the system. The tubing is then covered with a stabilizing material like Therma-Floor, a gypsum underlayment designed to pour over hot tubes, encasing the water tubes or electric cables in a crack-resistant, noncombustible gypsum.

boiler to reheat the water and complete another circuit (Figure 2.16). Radiators are positioned low to the floor and near window openings. This placement eliminates drafts, because as hot air rises and cold air sinks, the radiators reheat the cold air, thus keeping floors warm (Figure 2.17). Unlike forced-air systems, radiant heating does not control humidity.

Electric radiant heating systems have lower installation costs than boiler-fed systems, mainly because electricity is already planned for in the construction. However, the operating costs maybe expensive in regions with long winters and high electricity costs. Electrical radiant heating can be distributed through radiators, laid in ceilings, or embedded under floors.

Radiant floor heating systems utilize a series of pipes or electrical conduit embedded under the finished floor that is heated with either hot water or electricity (Figure 2.18). Consideration for the expansion and contraction of wood substrates must be taken into account to prevent floors from buckling. The most effective radiant floor system is one installed directly into the concrete slab. Radiant floor heating systems are more energy efficient than others and maintain an even distribution of heat, because warm air rises from the floor toward the ceiling at a constant rate.

FIREPLACES, AND WOOD-BURNING AND GAS STOVES

Fireplaces are popular in all regions of the country regardless of how cold the winters are. Nowadays, even houses built in warm southern locations feature a fireplace to take the chill out of damp winter air. Fireplaces are designed to fit into traditional-style homes as well as contemporary ones. Regardless of style, these fixtures all have the same components, and strict building codes regulate fireplace construction for obvious safety reasons (Figure 2.19). A *chimney* connects the *firebox* (where the

FIGURE 2.19 The contemporary design of this fireplace meets fire safety codes with its granite surround, chimney, and hearth.

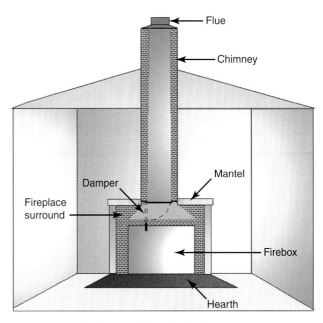

FIGURE 2.20 The inner workings of a fireplace.

fire is built) to an opening in the roof. Chimneys can be placed inside or outside the home, as long as they extend above the roofline far enough to keep roofing materials away from flying embers. The height of the chimney stack above the roofline is calculated to provide the most "draw" of oxygen for better fire combustion.

The chimney, hearth, firebox, and surround must be made of noncombustible materials like concrete, brick, stone, or CMUs. The *flue* lining the chimney must also be made from noncombustible materials—usually fireproof clay, stainless steel, or brick. The chimney is fitted with a *damper* that opens and closes the flue to regulate ventilation and allows smoke to escape through the chimney (Figure 2.20). The National Fire Protection Association (NFPA; www.nfpa.org) sets dimensions for the size of the hearth and the length it must extend from the firebox; however, municipal codes regulate how close the firebox can be to *combustible* materials like a wooden *mantel* or surround.

When advising clients, it is important to point out that conventional fireplaces are not energy efficient, because as much as 80% to 90% of the heat it produces escapes through the chimney. An alternative to traditional fireplaces, advanced combustion fireplaces increase output efficiency by 50%. Combustion fireplaces look similar to conventional ones and are equipped with vents, ceramic glass doors, and a fan. Vents draw in air from inside the room, and it is heated and recirculated back into the room. The fan aids in better combustion, producing more heat.

Like fireplaces, wood and gas stoves are regulated by strict codes that limit where they are placed inside the home relative to combustible materials. Zero-clearance stoves are allowed, as long as they are positioned against walls treated with noncombustible materials like stone, brick, tile, or fire-rated gypsum. Hearths must be made from noncombustible materials and should comply with municipal codes regarding size and position in front of the firebox (Figure 2.21). Every precaution must be taken to ensure the stove is installed to meet the manufacturer's guidelines and current fire codes protecting against house fires. Some municipalities require newly installed stoves to be inspected by the local fire department.

FIGURE 2.21 A wood-burning stove is angled in a corner with walls made of gypsum plaster and it sits on a terracotta tile floor. Both the wall and floor materials are noncombustible.

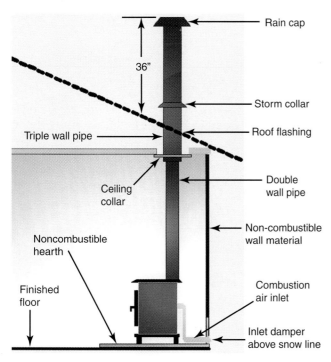

FIGURE 2.22 Component parts for a wood-burning or pellet stove.

Rain cap

36"

Storm collar

Roof flashing

Triple wall pipe

Ceiling collar

Double wall pipe

Noncombustible hearth

Non-combustible wall material

Finished floor

Combustion air inlet

Inlet damper above snow line

In 1990, the *Environmental Protection Agency* (www.epa.gov) set guidelines for making wood-burning and gas stoves more efficient to encourage their use as alternative heating sources. The newer model stoves burn fuel—wood, gas, propane, or wood pellets—more efficiently with less air pollution. Wood-burning and gas stoves are made from heat-conducting materials like cast iron, soapstone, and porcelain. Essentially, stoves work on the physics of radiant heat. Airtight stoves allow for hot fires that heat the surface of the stove, which in turn warms the surrounding air. Convection stoves work on a similar principle, except the firebox is surrounded by another layer of material to create an airspace between the radiant surface and the outer surface of the stove. Air circulating over the radiant surface of the inner layer heats faster, and the warmth is distributed into the room by blowers. By design, stoves are more efficient than fireplaces because most of the heat is radiated back into the room rather than through the chimney and to the outdoors (Figure 2.22). Chimneys for fireplaces and wood-burning stoves must have regular cleanings to remove the buildup of creosote in the flue; without regular maintenance, creosote can ignite and start a house fire.

Plumbing Systems

Residential plumbing systems are planned before the foundation is laid. In most areas, underground pipes for water and sewer connect to a municipal system, and are then brought up through openings in the foundation for direct access by a plumber (Figure 2.23). Two-by-six studs are used to frame *wet walls* where plumbing pipes and *waste stacks* are located, and a series of pipes is distributed to the wet areas of the house, like the kitchen, bathroom, and laundry room.

For rural settings or areas without a municipal system, a septic and well system is needed. A septic system uses a holding tank for effluent, or waste water, which is treated with natural bacteria. This filtered water is then distributed back into the ground through a leach field. The septic tank must be professionally pumped out every couple years or so, depending upon the capacity of the holding tank. Well

CAUTION

Plumbing Codes

Plumbing and sanitation codes protect the health of all residents of a community and are regulated by the municipality where the home is constructed. *International Plumbing Codes (IPC)* regulate the installation of water and sanitation pipes to ensure that the water brought into the home is safe for drinking and washing, and is removed in a manner that keeps noxious sewer gases from entering the home or contaminating the environment.

FIGURE 2.23 Waste stacks are already set in place between the stud walls as a plumber works on the copper piping.

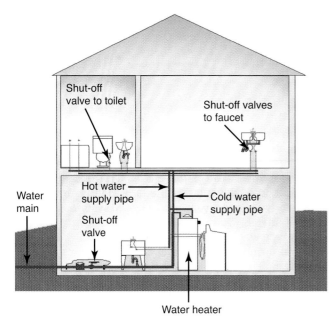

FIGURE 2.24 Cold and hot supply pipes fed from the main water line and hot-water tank are distributed throughout a two-story home.

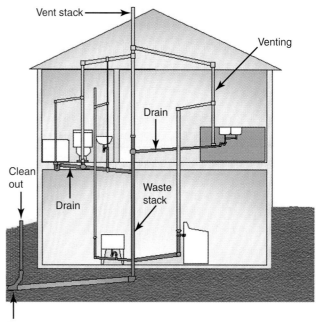

FIGURE 2.25 The plumbing system in residential construction relies on a series of pipes run through stud walls and floor joists. In a common plumbing system for a multistory home, soil and waste stacks connect to a vent stack rising through the roof. Hot water is generated from either the home's boiler system or from a hot-water heater located in the mechanical or utility space of the house.

systems are dug deep enough to reach groundwater that is either *potable*, or made potable through a filtration and purification system. A well requires a pump to operate, and homes with this type of system are often equipped with an auxiliary generator to keep the pump running in case of power failure.

PIPES AND DRAINS

Regardless of whether a home has a municipal water and sewage system or well and septic, the components of residential plumbing rely on a series of supply pipes (those bringing hot and cold water to a faucet), and drain and waste pipes to remove wastes (Figures 2.24 and 2.25). Fresh water is fed through either copper pipes or flexible *polyvinylchloride (PVC)* piping using pressure. It is important to note that some municipalities do not allow the use of PVC pipe and, as with any construction project, municipal codes must be followed. The main water supply coming into the home is equipped with an emergency shutoff *valve* in case of flooding resulting from broken pipes. Furthermore, the locations of water supply lines attached to sinks and toilets have additional shutoff valves that cut the flow of water to that specific fixture if necessary.

Waste water from lavatories or sinks, tubs, showers, and laundry rooms empty through drainage pipes connected to the main waste stack and is then distributed into the sewer system. To comply with sanitation codes, toilets must be directly connected to a *soil stack*. Waste and soil stacks are sloped using gravity to drain properly. The recognizable elbow bend in drain pipes is designed to trap water, creating a seal against poisonous *sewer gases* to prevent these noxious fumes from seeping into the house (Figure 2.26). Water contained in the toilet bowl acts similarly, keeping sewer gas from infiltrating the home. In addition, *vent stacks* extending through the roof connects to the waste and soil stacks channeling sewer gases out of the house. The venting maintains sufficient pressure to keep water flowing through the elbows of the drains (Figure 2.27).

FIGURE 2.26 A worker attaches a PVC drain into the waste stack in this new home.

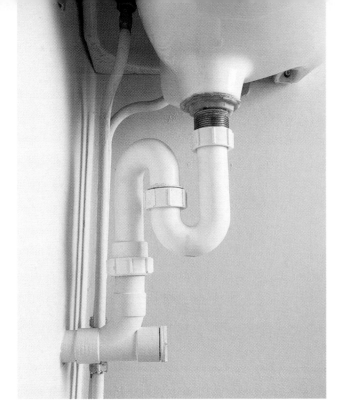

FIGURE 2.27 The elbow bend in the drain pipe traps water to keep sewer gases from entering the home.

LEED the Way

Water Heaters

Water heating systems earn LEED credits if the appliance meets the requirements for energy efficiency or their installation cuts down on energy waste. By reducing the distance between the heater and faucet, and by insulating the tank, points are given relative to energy savings. Eligible systems include both storage types and tankless types. Storage-type water heaters must be insulated with a minimum R4 factor (the *R factor* indicates the material's ability to resist heat transfer; the higher the number, the more efficient the system). Tankless systems provide hot water on demand. Heating water only when it is needed reduces energy consumption and potential heat loss compared with storage systems. For a complete comparison of hot-water heating systems, check out the *Consumer Guide to Home Energy Savings: Water Heaters*, published through the American Council for an Energy-Efficient Economy (www.aceee.org).

PLANNING FOR PLUMBING

In residential planning, it is more economical for the plumbing to be in the same general location. For example, having an upstairs bathroom located directly above a downstairs bathroom allows the two rooms to share waste and soil stacks. Furthermore, the fewer water pipes running through a home, the less damage done in case of leaks or broken connections. Areas for running plumbing pipes include the obvious—kitchens, bathrooms, and laundry rooms—but also include running pipes to specialty fixtures like icemakers in refrigerators, pot fillers on stoves, bidets in bathrooms, and specialized laundry sinks. Locations for plumbing appear on floor plans for new construction identified by architectural plan symbols for sinks, tubs, showers, and water-using appliances. It may be necessary to relocate plumbing access points based on client needs. An interior designer involved in remodeling projects needs to be familiar with how plumbing systems work to make any necessary changes, alterations, or improvements to an existing structure.

FIGURE 2.28 This cutaway view shows how copper supply pipes for hot and cold water are fitted to a faucet on the finished wall.

VALVES, FITTINGS, AND FAUCETS

A valve controls the flow of water through a pipe from its source to its destination. In simple terms, a faucet acts as a valve turning on or off the water supply to a sink, tub, or shower. Fittings make the connection between the pipes, which are embedded in walls and floors to the faucets. Specifications for faucets, including showers, tub fillers, and sinks, must be determined during the early stages of a construction project and given to the builder or plumber to ensure compatibility between pipe and valve fittings (Figure 2.28).

The shower valve is one of the most important considerations for any home design project. Valves modulate the mixing of hot and cold

Active solar energy systems rely on solar collector panels that store energy from the sun until it is ready to be used later. Solar energy can supply enough power for heating and air-conditioning systems and hot-water heaters, and can run home appliances and lights. Although the technology for solar energy has existed since the first oil energy crisis during the 1970s, even today the widespread use of solar energy is not what it could be. One factor may be that these systems have a higher initial cost for purchasing equipment and installing the system, but after these systems are operational, the savings are worth the investment.

A roof-mounted solar heating panel provides the energy needed for operating a hot-water heater.

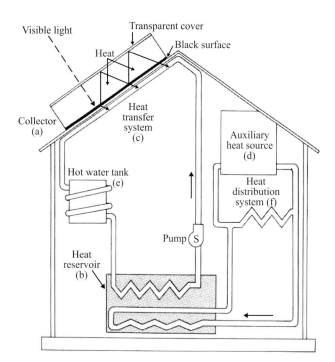

This diagram shows how a solar collector on the roof is used to transfer and store heat for later consumption.

water under controlled pressure, and incorporate scald protectors in case there is an unexpected reduction in cold-water pressure. Suppliers of shower valves and faucets offer thermostatic controls that regulate water temperature. A dial is set to the desired temperature and, when the faucet is turned on, the preset temperature is maintained for as long as the water is running. These valves should be purchased from the same manufacturer as the shower faucet, because they are sized to work with specific products. More information on selecting faucets and plumbing fixtures is presented in Chapter 3.

resources
informative websites

American Council for Energy Efficient Economy: www.aceee.org

American Wind Energy Association: www.awea.org

Energy Star: www.energystar.gov

Environmental Protection Agency: www.epa.gov

National Fire Protection Agency: www.nfpa.org

Partnership for Advancing Housing Technology: www.pathnet.org

Interior Finish-Out for Residential Construction

Enclosing the Envelope

Construction on finishing the interior of a residential project involves many steps before the home is ready for occupancy. Coordination among the contractor, architect, and interior designer to ensure that all the interior finish materials have been specified, ordered, and delivered to the job site when needed is a critical part of the interior finish-out phase. No one wants to see workers standing around with nothing to do because they are waiting for doors or windows to be delivered. More important, the *finish carpentry* work cannot begin until the building envelope is completely sealed against inclement weather.

Exterior doors and windows with related hardware are installed to protect the space before work begins on the inside. Then, stud walls and ceiling joists are enclosed, light fixtures are installed, and roughed-in staircases are given their final shape (Figure 3.1). Kitchens and bathrooms take final form with the installation of plumbing fixtures and faucets, cabinets and vanities, and finishes for floors and walls. Work is done along with the rest of the finished carpentry, like installing wall paneling, door and window trims, molding details, and built-ins.

The dining and kitchen in this arts and crafts-style home feature finished carpentry work with custom paneling, ceiling beams, and matching door and window casings.

FIGURE 3.1 Studs define the shape of the interior space for walls and ceilings, and are ready for installing the electrical wiring and lighting, HVAC ductwork, and plumbing lines.

Wall Assemblies

All walls throughout the interior of wood frame construction must be covered to conceal electrical wires, plumbing pipes, and framing studs. Deciding which walls are covered with *wallboard*, wood paneling, tiles, or something else is done well before construction begins. Walls finished in tile or stone may need special stud spacing to accommodate their weight along with important backing materials. In addition, walls where glass or glass block will go are framed differently than those covered with *gypsum* wallboard or wood paneling.

GYPSUM WALLBOARD

One of the most common methods of enclosing stud walls is using wallboard products made from gypsum. Often referred to as *drywall* or sheetrock, gypsum wallboard is made from compressed gypsum sandwiched between layers of thick paper. Gypsum wallboard is available in different thicknesses and, for most residential projects, thicknesses of five-eighths of an inch are applied to walls, and one-half of an inch are used on ceilings. Sheets are produced in 4 by 8-, 4 by 12-, and 4 by 16-foot dimensions; component sizes are based on typical wall length modules and 8-foot ceiling heights. Gypsum wallboard panels are attached to stud walls and ceiling joists with drywall screws. Holes are cut out for accessing junction boxes for walls, ceiling outlets, and switches (Figure 3.2).

For concrete walls, drywall is attached using adhesives or metal *furring channels*. Adhesives are spread onto the concrete wall and drywall is applied directly to the concrete (Figure 3.3). In regions with high humidity, metal furring channels leave a space between the drywall and the concrete. This air space keeps the drywall from coming into contact with concrete blocks that might sweat. Furring channels

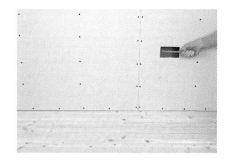

FIGURE 3.2 Cable for data and communications ports is pulled through an opening in the drywall.

LEARN More

Gypsum Wallboard

The core material used to make gypsum wallboard is made from a slurry of powdered calcium sulfate dehydrate mixed with binders that, when dried, produces a rock-hard material. The gypsum core is pressed to thicknesses ranging from one-quarter inch to five-eighths inch and is fabricated into panels size 4 by 8 feet, 4 by 10 feet, and 4 by 12 feet. Calcium sulfate dehydrate, or gypsum, is a mineral that is inherently noncombustible, providing fire resistive protection for interior wall applications. Moreover, gypsum wallboard also controls sound between spaces, and special gypsum wallboard can be specified to control moisture in areas like kitchens, bathrooms, and laundry areas. To learn more, visit the Gypsum Association's website at www.gypsum.org.

FIGURE 3.3 A worker uses a plaster trowel to smear pink drywall adhesive on a concrete wall.

GO GREEN Insulation

International Energy Conservation code requires that residences be insulated against heat loss in attics, ceilings, walls, floors, crawl spaces, and basements. Insulation placed inside the stud walls provides thermal control and acts as a sound barrier. The insulation creates an air-tight stud cavity that prevents air infiltration, which leads to heat loss or gain. There is a variety of insulation materials on the market for residential construction; the most familiar is the pink batt type available packaged in rolls. These rolls are manufactured in 16-inch and 24-inch widths to accommodate conventional stud spacing. This roll type of insulation is made from fiberglass, although new ecofriendly alternatives made from cotton batting have recently been introduced. Cotton is a renewable source, which makes it the "green" choice, but cotton is naturally combustible and must be treated with a flame retardant to meet building codes. Additional insulation products on the market for residential applications include rigid foam sheets and a spray-on foam that fills all the gaps to provide an air-tight seal. Insulation is measured in R values determined by its capacity to reduce thermal loss or gain. Insulation with a higher R value provides greater thermal protection than that with a lower value. For more information, visit the International Code Council website at www.iccsafe.org.

Insulation fit between studs provides thermal and acoustical control.

are anchored to the concrete with masonry screws. The wallboard is then screwed onto the furring channels every 16 inches (Figure 3.4). For all applications, care must be taken not to tear the paper on the wallboard.

Tape and bedding joints between drywall sections prepare the wall for receiving the final finish material. A thin layer of *joint compound* is applied over the seams using a 4-inch putty knife. A specially made paper tape is then applied and allowed to dry before a second coat of joint compound is applied over and beyond the edges of the tape (Figure 3.5). After this second coat of joint compound dries, the joints receive a third and final finish coat of joint compound, during which skilled workers feather the edges to hide any remaining indication of the seam. After the tape and bed work has thoroughly dried, the joints are then "sanded" with a wet sponge, giving the wall a smooth surface. *Blue board*, a gypsum core product sandwiched between layers of blue paper, does not need to be taped and bedded. After the blue board is screwed in place, the wall surface is covered with a thin layer of plaster. When dry, the plastered wall is ready for painting or wallpapering.

FIGURE 3.4 A worker uses a power tool to attach drywall to metal furring channels.

The process of finishing wallboard to a smooth, flat surface is not uniformly adopted throughout the country. In fact, some regions of the southwest, influenced by the thick adobe walls of old houses, prefer a more textured surface. To achieve a rough texture, the prepared drywall is covered with wall plaster using a large trowel, spread thinly in some areas and thick in others (Figure 3.6). When dry, the plaster is ready for paint. This same texturing process can be applied to concrete walls, sparing the expense of applying wallboard.

After drywall has been taped and bedded, the absorbent paper must be coated with a *primer* before painting. If the drywall is not primed, the paper will soak up the paint and the color will appear splotchy or uneven. Moreover, drywall primer has a high pigment content designed to hide taped and bedded areas. After the primer has dried, the finish paint is applied in one or more coats, resulting in a thickness measured in millimeters according to manufacturers' specifications. When using dark colors, the primer should be tinted with the base color to avoid a chalky look (Figure 3.7).

FIGURE 3.5 Joints between installed drywall are taped and bedded, giving the wall a smooth surf.

FIGURE 3.6 Hand-applied plaster gives drywall a rough surface texture.

Nontoxic paints made with natural pigments taken from plants and minerals mixed with milk protein, plant oils, or bee's wax binders were used to paint the interiors and exteriors of colonial homes. Unlike a lot of paints today, these all-natural products were not harmful to the environment. The current environmental crisis has led manufacturers of paints and coatings to develop new products without harmful *volatile organic compounds* (VOCs) in an effort to reduce greenhouse gases. New products recently introduced to the market include paints and coatings with zero VOCs and those with low VOCs, earning the Green Seal stamp of approval. Green Seal's mission is to achieve a more sustainable world by promoting environmentally responsible production, purchasing, and products. A product with the Green Seal logo ensures it meets rigorous, science-based environmental leadership standards. For further information, visit the organization's website at www.greenseal.org.

The Green Seal logo is restricted for use on those products that meet high standards for environmental concerns.

GLASS BLOCK WALLS

Glass block walls became popular during the Art Deco period of the 1920s and 1930s when inexpensive machine-made blocks were introduced. Before mass production, glass blocks were handmade. Two pieces of glass are vacuum sealed, leaving air space between them. The air space provides thermal insulation, making glass block suitable for both interior and exterior walls, and it controls *sound transmission*. Glass blocks are installed in courses; the first course is laid on a sill plate and each block is mortared into place. The next course is stacked using plastic spacers to leave room for mortar and metal reinforcing rods that stabilize the wall as it is built higher (Figure 3.8). Glass blocks are 4 by 4, 6 by 6, 8 by 8, and 12 by 12 inches, and there are curved blocks for turning corners, end caps for freestanding walls, and angled blocks for 90-degree corners. The availability of multiple colors and textures inspires designers to create bold architectural statements with glass block that emphasize the translucency, texture, and visual impact of the glass with special lighting (Figure 3.9).

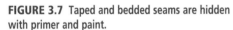

FIGURE 3.7 Taped and bedded seams are hidden with primer and paint.

Ceilings

During the framing of the house, ceiling joists are arranged according to the desired shape of the finished ceiling. Breaking away from tiresome flat ceilings, cathedral ceilings peak toward the roofline, sloped ceilings might mimic the lines of a shed roof, and recessed or tray ceilings project into attic spaces above. These shapes give rooms the feeling they are bigger than they actually are and add interest to the interior.

FIGURE 3.8 Reinforcing rods placed between each coursing of glass block and a cement mortar ensures stable wall construction. Plastic spacers keep a uniform alignment.

Ceilings are insulated to reduce heat loss from the interior. As warm air rises, it escapes through the ceiling, into the attic, and out the roof. Insulation is either laid in or blown in between the finished ceiling and the attic space above. A *batt insulation* product with a foil layer adds to the efficiency of regular insulation by 30% (Figure 3.10).

Gypsum wallboard, wood, metal, and fiberboard are examples of materials used to enclose ceiling joists and are chosen for a specific aesthetic. Wood ceilings offer a warm, natural look, although they are more expensive than gypsum wallboard. Wallboard panels are attached to ceiling joists with screws placed every 12 inches to ensure that the wallboard does not deflect, or sag, over time. Wallboard

FIGURE 3.9 A dramatic centerpiece between the entry hall, living room, and dining area, this curved glass block wall is enhanced with special lighting.

FIGURE 3.10 Steps for maximizing control over heat gain or loss in this attic space include placing pink blanket insulation between the studs, attaching foil sheathing over the insulation, and applying a layer of thermal-check plasterboard.

ceilings are taped and bedded in the same way as walls, and are then primed and painted (Figure 3.11). Wood ceilings directly applied to ceiling joists use lightweight wooden slats, or *engineered wood* like Glu-Lam, that are fastened together into panel sections. These sections are screwed to ceiling joists and the screws are sunk and covered with a wood plug to give the ceiling a more finished look (Figure 3.12).

A suspended ceiling has a lightweight aluminum grid hung from the structural ceiling joists with hanging wires. The grid supports drop-in tiles made from lightweight woods and metals, glass, plastic, or acoustical fiberboard. Acoustical fiberboard was first used in homes designed during the post-World War II housing boom (Figure 3.13). Designed as an economical and easy-to-install ceiling treatment for controlling sound in noisy kitchens and basement recreation rooms, fiberboard tiles

FIGURE 3.11 After wallboard is taped and bedded, a primer is applied to seal the paper and prepare the surface for paint.

CAUTION

Code Check: Ceiling Heights

Building codes require that ceiling heights for R3 occupancies be a minimum of 7 feet 6 inches for at least 50% of the total room area, and no part of the ceiling can be less than 5 feet from the finished floor.

FIGURE 3.12 This ceiling is finished with Glu-Lam planks (prefabricated panels made from glued wood by-products laminated under heat and pressure).

were made with cancer-causing asbestos fibers and *formaldehyde*. Today's products are safer, made without asbestos or harmful volatile organic compounds (VOCs). An important note to consider when specifying suspended ceiling systems is that framing must be planned in advance to ensure the height of the finished ceiling meets code.

FIGURE 3.13 Recessed acoustical ceiling tiles create visual interest in this recreation room.

LEED the Way

Rapidly Renewable Materials

LEED offers one point toward earning certification when 2.5% of the total building materials and products come from plants that can be harvested in cycles of 10 years or less. These substances include cork, plantation-harvested woods, and bamboo. Bamboo regenerates, quickly replenishing supplies, whereas cork and plantation woods are harvested in 10-year cycles. For more information on green building practices and sources, visit the Northwest EcoBuilding Guild website at www.ecobuild.org.

FIGURE 3.14 Metal ceiling tiles enhance the period look of this kitchen and eating area.

In areas where a dropped ceiling might be too low, manufacturers of suspended tile systems offer tiles that can be directly applied to OSB or plywood *substrates*.

Victorian homes featured kitchens with metal paneled ceilings as fireproofing against cooking fires. Panels were made from pressed tin and stamped with an assortment of patterns ranging from simple geometrics to fancy floral designs. Modern reproductions of these metal ceiling tiles are available with new designs and enamel coatings in an array of colors (Figure 3.14). Metal ceiling tiles are installed using either a suspended ceiling system or are applied directly to a substrate of plywood or OSB sheathing. Modern metal tiles are made from recycled aluminum, which is lightweight and easily formed into patterned sheets. Manufacturers offer other options, including tiles that look like metal but are made from recycled plastics.

GO GREEN   Carpeting

Ecofriendly carpeting does not rely on petroleum-based fibers or backing materials, and it is manufactured with zero- or low-VOC adhesives. Carpet made from wool fibers is the green choice for new homes. Wool is a naturally sustainable fiber that is sheared from sheep without causing any harm to the animal. Sheep's wool is a protein fiber that is naturally fire resistive, soft but durable, and can be dyed. Ensuring that wool carpeting is installed with zero- or low-VOC adhesives over a pad made from recycled materials contributes toward a healthy home environment. For the full Green Seal report on carpets, visit the following link: www.greenseal.org/resources/reports/CGR_carpet.pdf.

Flooring

Floor finish materials like carpeting, wood, ceramic or stone tiles, vinyl, cork, synthetic rubber, and *linoleum* are floors specified for their aesthetic, performance, and cost. Most finish flooring materials can be laid on top of a poured concrete slab or *subfloor* sheathing materials without special preparation; however, certain materials like stone or tile require preplanning during the framing stages of the house because of their weight (refer back to Figure 1.36).

Flooring materials are a contributing factor toward poor indoor air quality, because many products and adhesives used in their installation emit harmful toxins. Growing awareness of the damage caused to the environment by these toxins has led manufacturers of flooring materials to eliminate or reduce VOCs during production processes. When considering going green,

FIGURE 3.15 Sheet carpet is laid wall to wall.

ecofriendly materials include products manufactured without carcinogens like formaldehyde, and those made from sustainable or recycled materials.

CARPET

Residential carpet is available in 6-, 12-, and 15-foot widths packaged on large rolls (Figure 3.15). *Broadloom carpet* is sized to minimize the amount of seams required when installing the material in large spaces. Carpet is installed over a pad specified for its cushioning properties. Carpet pads extend the life and performance of the carpet, and provide a buffer between the carpet backing and OSB or plywood substrate material. Without cushioning, carpet backing materials will break down and the carpeting will wear unevenly and have to be replaced. Broadloom carpeting is installed using a carpet tack, a narrow strip fitted with upward protruding tacks nailed to the subfloor along the perimeter of the room. The edge of the carpeting is turned under onto the tack strip and a mallet is used to secure the carpet to the tacks. The carpet is then stretched across the room to another tack strip; over the life of the carpet, it may need to be restretched to keep the material taut.

HARDWOOD AND LAMINATE FLOORING

Wood flooring is specified in residential projects for the warmth it gives to any interior (Figure 3.16). Hardwoods like oak, maple, cherry, hickory, and birch, and softwoods like pine or Douglas fir are milled into thicknesses of one-half-inch, five-eighths-inch, or three-quarter-inch solid wood strips and planks. The thicker the wood, the more often the floor can be sanded to remove scratches and dents, and refinished over the lifetime of the floor. Solid *wood plank flooring* and *strip flooring* is installed with interlocking *tongue-and-groove* joints that are milled into each board length. Tongue-and-groove joints allow the wood to expand and contract with changes in humidity. The wood lengths are laid with staggered joints, and a half-inch *expansion gap* is left between the wood strips and the wall (Figure 3.17). This gap is then hidden by a baseboard or piece of trim, which still allows the wood to expand and contract (Figure 3.18).

Random lengths
staggered joints

Uniform lengths
staggered joints

FIGURE 3.17 Variable lengths of strip floor-ing are laid in regulated or random patterns with staggered joints.

FIGURE 3.16 Wood flooring in cherry is laid in narrow strips in this breakfast room.

Laminate wood flooring has the look of solid hardwood but is an engineered product manufactured with increased dimensional stability, and it guards against changes in humidity, which cause wood to warp (Figure 3.19). This engineered flooring product is made by laminating wood *veneer* using glues and pressure over a *high-density fiberboard* core to form panel segments. The out-ermost layer is a clear protective coating designed to withstand nicks and scratches, and it protects the wood veneer from water damage. A resin-coated backing protects the laminate floor from moisture, making it suitable for basements and concrete subfloors.

Plastic laminate flooring simulates real wood with a high-quality digital image sealed into layers of plastic. Both engineered laminate flooring and plastic laminate flooring locks together easily with tongue-and-groove joints. These finishing materials

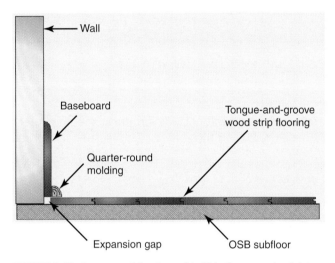

FIGURE 3.18 Quarter molding is used to hide the expansion joints between wood floor and wall.

LEARN More

Bamboo

Bamboo is a fast-growing plant typically grown in the Far East. Because bamboo is a plant and not a tree, it is a sustainable material as, even after cutting, it continues to grow. Flooring manufacturers conscientious about the environment have adopted bamboo as their newest ma-terial for making laminate flooring. The bamboo is har-vested and the shoots are flattened into sheets. Moisture is removed and the flattened stalk is laminated onto a wood substrate using glues and pressure. The beauty of the distinctive growth knots of the bamboo is evident in the finished product and is visible even after staining. Laminated bamboo flooring is engineered to a hardness equal to or exceeding maple and oak. In addition, bam-boo is not just used as flooring material, but also as wall paneling and a ceiling treatment. For further information and updates on the advances of bamboo in building con-struction, visit the Environmental Bamboo Foundation website at www.bamboocentral.org.

A view of a bamboo grove in Japan.

FIGURE 3.19 Laminate flooring is a durable choice for the floor in this beach house.

are also installed directly onto the substrate with adhesives. To reduce the sound of hard-soled shoes walking on hardwood floors, both laminate and hardwood floors benefit from a thin *underlayment* of felt, cork, or polyethylene padding placed on top of the subfloor to absorb sound.

CERAMIC AND QUARRY TILE

Ceramic tile is made from fine clay and water. Cut tiles are fired in a kiln to give the material its strength and durability. Before firing, the tile may be glazed to add color and sheen to its surface. Surface glazes vary from shiny and smooth, to those with a low sheen, *matte glaze,* or crystallized finish. Quarry tiles are made from a combination of fine clay, shale, and other earthen materials that give the tile its color, typically recognized by the terra-cotta or rust-red shades. Quarry tiles are usually left unglazed, and after firing are superresistive to water.

Ceramic and quarry tiles are good choices for floors and wall coverings in wet areas like kitchens, bathrooms, and laundry rooms, because they are nonabsorbent and easy to clean. Water resistant tiles are applied to walls, countertops, backsplashes, and floors. Tile is rated according to its water absorption rate. Ceramic *glazed tiles* have a water absorption rate of 18% or lower and *unglazed tiles* have a rate of 3% or less.

The strength and durability of tile makes it a suitable flooring material for hallways, entryways, and mudrooms (Figure 3.20). Moreover, tile is inherently cool,

FIGURE 3.20 An entry hall covered with terra-cotta tiles provides rugged durability for high-traffic areas.

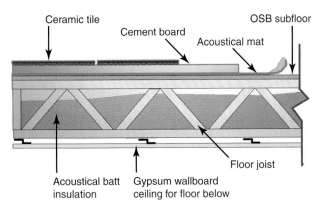

Ceramic tile
Cement board
OSB subfloor
Acoustical mat
Floor joist
Acoustical batt insulation
Gypsum wallboard ceiling for floor below

FIGURE 3.21 For greater sound control, ceramic tiles and cement board are laid over acoustical underlayment, and floor joists are filled with batt insulation.

which makes it a popular flooring material for homes in warm climates. Although tile is a practical solution for areas needing a heavy-duty floor material, there are things to consider when specifying tile in residential projects.

A tile floor must be planned for during the framing stages of the house. If tile is to be installed on walls or floors, studs and floor joists should not exceed 16 inches on center and, depending on the combined weight of tile and backing materials, floor joists may need to be spaced 10 inches on center to prevent deflections in the floor that could crack tiles.

Tile is applied on top of an underlayment of *cement board*, sometimes called *hardy backer*. The cement board is attached to the subfloor with screws to provide a stiff and sturdy substrate and to ensure a strong adhesive bond between tile and floor. The rigid cement board prevents tiles from cracking under the applied weight of furniture and people. The hard, resonant surface of tile floors amplifies sounds. Applying an acoustical mat over the subfloor material but layered beneath the cement board will help to control sound (Figure 3.21). In addition, acoustical batt insulation added between floor joists will keep sound from traveling between floors.

Factoring in the subfloor, underlayment, and tile itself, tiled floors will have an overall thickness greater than carpet, hardwood, or resilient flooring. Considering how different flooring materials transition between adjoining spaces must be planned in advance. Variable thicknesses of OSB or plywood underlayment are used to maintain a continuous level between rooms with different flooring materials.

Ceramic tiles are spaced anywhere from one-quarter to three-eighths inches apart and are laid with a mortar adhesive. After the tiles are laid, the floor cannot be walked on until the mortar sets, usually after 24 hours. The gaps between tiles are then filled in with *grout* (Figure 3.22). Glazed tiles and unglazed tiles require different types of grout. Sanded grout is applied over unglazed tiles; nonsanded grout, which will not scratch the surface, is used over glazed tiles. After the grout cures, it keeps water from seeping down into the substrate material. A liquid sealant can be brushed over the grout lines for better performance and longer wear. Grout colors are chosen to complement, contrast, or blend with the color of the tiles, and have antimicrobial properties, are stain resistant, and resist shrinkage.

FIGURE 3.22 The last step in installing ceramic tile floors is the application of grout. Its pastelike consistency is pushed into the gaps between tiles; excess grout is wiped clean and the grout is left to dry.

FIGURE 3.23 Black-and-white vinyl tiles arranged in a checkerboard pattern are set with floor adhesive.

RESILIENT FLOORING

Resilient flooring materials are pliant, which makes them good choices for areas where cushion comfort is needed. Manufactured for resisting stains and suitable for high-traffic areas, resilient flooring products are made from man-made materials like vinyl and synthetic rubber, and natural materials such as rubber, cork, and linoleum. These materials protect the subfloor from water seepage, are easy to maintain, and are economically priced. Resilient flooring materials are installed over the subfloor using adhesives. Resilient flooring should not be installed on concrete floors below grade because ground moisture will loosen the adhesive bond.

Vinyl flooring is manufactured in rolled sheets or tiles and is available in a variety of colors, patterns, and textures (Figure 3.23). Vinyl flooring is made from a core of polyvinyl chloride and plasticizing solvents. The core can be a solid color throughout or covered with a digital image of just about anything from granite, wood, and even tile. For better wear and durability, the digital image is treated with a protective finish. Both solid-core and printed vinyl are backed with felt or heavy paper.

Vinyl is a petroleum-based product, and many of the adhesives used to install it are comprised of formaldehyde. In addition, the emissions from the chemicals used to produce and install vinyl flooring contributes to poor indoor air quality. Awareness of the hazards of poor indoor air quality has led manufacturers to develop formaldehyde-free products and adhesives, making these products greener and healthier for residential environments.

Specifying flooring materials made from natural ingredients are better choices for green-minded designers. Linoleum is made from linseed oil, pine tree resin, wood flour, and natural pigments. These organic materials are formed into sheets or tiles and are backed with natural jute fibers. Linoleum was first introduced during the mid 19th century, but fell out of favor when vinyl flooring was introduced during the housing boom of the 1950s. Linoleum is available on rolls in 6-foot or 12-foot widths, or in 12-inch-square tiles. Because linoleum is made from natural materials, it is gaining popularity as a suitable replacement for vinyl (Figure 3.24).

The bark of the cork oak tree grown in areas of the Mediterranean is used to make cork flooring (Figure 3.25). The cork is harvested at least every 10 years, and waste by-products from the wine-making industry are used to make cork boards and flooring tiles. Cork flooring is produced by mixing ground-up cork (sometimes burned or scorched for coloring) with adhesives that is then formed into thin sheets under pressure. The surface of the cork is sealed with either a

FIGURE 3.24 Sheet linoleum is laid in colored sections on this basement floor.

wax or *urethane* coating. Maintaining green design practices, both linoleum and cork flooring should be installed using solvent-free or low-VOC adhesives. A note of caution, cork flooring is more susceptible to dents caused by heavy furniture.

FIGURE 3.25 Like any other resilient flooring tiles, cork tiles must be removed from their packaging a few days before installation to allow them to settle at room temperature. The tiles have naturally occurring shade variations, so sorting the tiles before installation helps distribute the colors evenly throughout the room. Adhesive is then rolled onto the substrate and allowed to dry until sticky to the touch, at which time the cork tiles are then laid.

Millwork

Woodworking details for staircases, ceiling moldings, wall paneling, baseboards, and door and window trim are part of the finish carpentry called *millwork*. Millwork details are the most visible sign of the quality of the construction inside the home. If the finish carpentry work is sloppy, the entire project looks second rate. Millwork details are either bench built, sometimes onsite, or purchased from lumberyard stock. In both cases, wood is milled or shaped into moldings and trims that are carefully scaled to fit the proportions of the room, taking into consideration ceiling heights and overall room dimensions.

STAIR DESIGN

Staircases fulfill both a practical function and an aesthetic one. More than just a way to get from one floor to the next, they are often designed as an important focal point between the main entry and living room (Figure 3.26). Considering the prominence of staircases in the home, they are designed to reflect the traditional character or contemporary style of the overall interior design scheme. Traditional styles might feature elaborate woodworking, turned handrails and *balusters*, and wooden steps. Contemporary-style staircases may not be wood at all, but may be formed from metal or glass, with modern-looking open *risers* and straight *rails*.

In case of emergency evacuations from situations like fire or smoke, staircases are usually the only way to get downstairs and outside quickly. Therefore, building codes are strict in regulating stair construction. Staircases cannot be too steep; each step must be the same height from one to the next. Keeping each riser the same height establishes an even rhythm that matches the normal gait of people going up or down so they do not trip and fall.

Components of staircase construction include risers, *treads, stringers,* balusters, and railings (Figure 3.27). The face, or front part of each step, is called a *riser,* and treads are what people put their feet on for each next step. Stringers are the structural supports for the risers and treads. Finish materials for risers and treads include wood, carpet, metal, stone, tile, or glass and can be used as long as the weight of the material was calculated into the structural support for the staircase. All materials covering treads must have a nonslip surface to prevent slips and falls. Balusters and the handrails they support must comply with municipal codes, which could restrict their design and installation. Glass stairs are made from tempered or safety glass and are prefabricated from specialty suppliers.

FIGURE 3.26 A winding staircase is framed and ready for completion in this new construction project.

Code Check: Stairs

The International Residential Building Code (IRC) is the model code that sets requirements for residential occupancies in the R3 group for stair construction. These codes state that riser heights must be a minimum of 4 inches and a maximum of 7¾ inches, and treads must have a minimum depth of 10 inches. Furthermore, staircases must be a minimum of 36 inches wide with handrails mounted on at least one side at a height of 34 to 38 inches above the finished floor.

Municipal codes may have stricter requirements than IRC; in fact, some towns require balusters to be arranged vertically and no farther than 4 inches apart. Vertically arranged balusters discourage children from climbing and falling over the side, and the spacing keeps curious children from sticking their head through the balusters and getting stuck. The most important rule in designing staircases is to check with the code enforcement officer for the city where the stairs will be built to be certain of all requirements.

Codes also determine what clearances are required as the staircase rises up to meet the next floor. Head clearance is the distance between the actual step and the ceiling above. A minimum head clearance of 6 feet 8 inches is typical. This provides enough space for most people to go up or down the stairs without hitting their head. The impact of this code determines

how much of the second story space is opened to the stairwell. Furthermore, all openings surrounding the stairwell must be blocked by either a wall, half wall, or railing to prevent someone from falling.

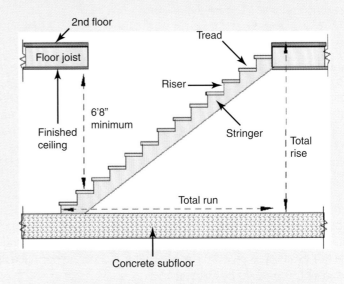

The required clearances and measurements in stair construction, which meet building codes, are marked.

The amount of floor space needed for a staircase is considerable; the length, width, and rise of the stairs obstruct the space beneath it, and often these areas are converted to storage space. However, there are many space-saving configurations that will help recover valuable floor space (Figure 3.28). A straight run of stairs is configured in a direct line without any turns from top to bottom. L-shaped staircases save floor space because the stairs are usually tucked into a corner of a room. By design, L-shaped staircases have landings at the 90-degree turn. Building codes allow

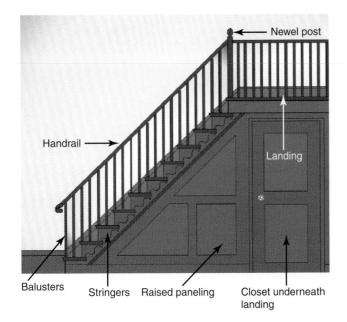

FIGURE 3.27 This drawing illustrates the millwork parts of a staircase.

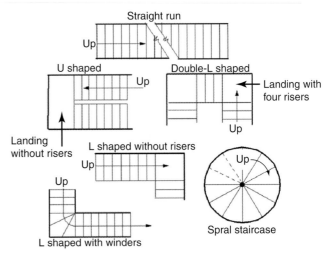

FIGURE 3.28 Drawings illustrate various staircase configurations.

at least one winder step, or a step that divides the landing with one riser, as long as the greater portion of the tread meets the minimum width requirement of 10 inches. U-shaped staircases are tightly configured and may or may not have risers in the landing. Spiral staircases are used in spaces too tight for conventional stairs, but getting furniture up or down a spiral staircase is somewhat problematic. Moreover, spiral staircases are not considered to be a safe means of exit, and for this reason should not be the only staircase in the home.

STAIR CALCULATIONS

Calculating the amount of risers and treads needed for building a staircase is relatively easy. First, the floor-to-floor height is measured from the anticipated finished floor of one story to the finished floor of the next. The floor-to-floor measurement is divided by a standard riser height of 7 inches. For example, if the floor-to-floor height is 8 feet, or 96 inches, then $96 \div 7 = 13.714$. The result is rounded up to the next whole number. In this example, the staircase would have 14 risers. The number of risers is then divided into the floor-to-floor height to determine a consistent measurement for the riser heights ($96 \div 14 = 6.85$). A staircase with riser heights set at 6.85 inches meets code requirements because it does not exceed 7¾ inches, but the staircase has a gradual incline and will take up more floor space when built. Using the same formula but changing the riser height to 7½ inches will end up with a steeper staircase than the one calculated for 7-inch risers. An important note to remember: If the floor-to-floor height is more than 12 feet, straight-run staircases are required by code to have at least one landing to break up the length of stairs (Figure 3.29).

INTERIOR PANELING

An expensive luxury for many clients, the high-brow beauty of custom-joined paneling adds warmth and richness to an interior. Custom-made wall paneling, whether covering the entire wall surface or only part of it, is made in sections and fitted together using traditional joinery techniques like tongue-and-groove connections. Solid lumber is milled into panels and joined to supporting frames called *stiles* and *rails*. The framing allows the panel to expand or shrink with changes in humidity (Figure 3.30). Instead of *raised paneling*, contemporary-style paneling is flush or flat, creating a smooth, uninterrupted surface, or is installed with narrow recesses between panels (called reveals) for interesting three-dimensional effects (Figure 3.31).

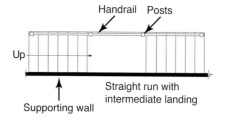

FIGURE 3.29 A straight run of stairs in spaces with ceilings of 12 feet or more must be broken by a landing to meet code.

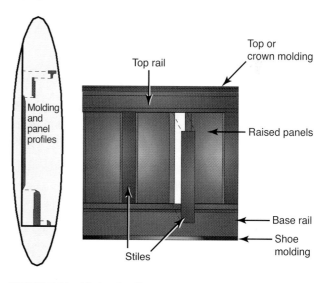

FIGURE 3.30 This drawing illustrates how wood panels and moldings are assembled in hand-joined millwork.

FIGURE 3.31 Flush wall paneling gives a contemporary look to this entry hall.

FIGURE 3.32 A narrow molding called a chair rail is placed between a panel of bead board and drywall. Wainscoting is topped off with a chair rail.

FIGURE 3.33 Wood battens are attached to an existing wall in preparation for applying wood paneling.

Wainscoting is wood paneling covering half or three-quarters of the wall surface. The stud walls where wainscoting will be placed are not fully covered with gypsum wallboard. Only the upper painted or wallpapered sections are covered with wallboard; the paneling is secured to bare studs. By installing wainscoting directly to the studs, the wall surface remains as one contiguous plane. A narrow strip of wood molding, called a *chair rail,* is attached where the paneling meets the wallboard, and baseboards cover the bottom nearest the floor (Figure 3.32). For renovation projects, paneling is secured to evenly spaced wood furring channels that are attached to the existing wallboard (Figure 3.33). Adding a chair rail with variable profile thicknesses makes the transition between the uneven surfaces.

Prefabricated wall paneling sold in 4-by-8-foot sheets is readily available for quick installation. Paneling made by laminating wood veneers over a core of *medium-density fiberboard* gives wall paneling added strength and dimensional stability that prevents the wood from warping and cracking. Thin wood veneers are cut from logs and bundled together. Using veneers from the same bundle or *flitch* ensures the wood graining, character marks, and color matches from one veneer to the other. The thin veneers are "sewn" together with a bead of glue, and heat is applied to mesh the seams into designed and matched sheets that are then glued to the core materials. Veneers are pattern matched to bring out the beauty of the grain (Figure 3.34). Mockup samples showing stain color, graining, and pattern match are shown to the architect or designer for approval before all the panels are laid out for installation.

MOLDINGS, BASEBOARDS, AND TRIMS

Molding designs used in the Victorian-style homes of the late 19th century once evoked quality, luxury, and wealth. Crown molding, picture molding, chair rails, and trim around doors and windows in these homes were fancily designed with curlicues, beads, flowers, and vines (Figure 3.35). By the early 20th century, superfluous ornamentation and details seemed too heavy for unencumbered modern tastes. In fact, the modern movement eliminated "decoration" from architecture and interior design, leaving interiors bare of moldings and trims. However, in today's housing market, traditional-style homes sell well, and interiors fitted with ceiling moldings, baseboards, and door and window trim are signs of quality.

Millwork details include crown molding (also called *cornice molding*), window and door trim, chair rails, and baseboards. Crown molding gets its name from its position at the top of the wall where it meets the ceiling line. In high-end residential construction, wood moldings and trims, like wall paneling, are usually bench built by a finish carpenter. However, *stock moldings,* available in 6- to12-foot lengths,

LEED the Way

Composite Wood Products

Specifying composite wood products over those milled from solid wood saves trees, because the composites—wood chips, wood fibers, and sawdust—come from the cutting, sawing, and milling of other wood products. This recycled wood, along with plantation timber, cuts down on the dependency of mature trees harvested for their size. Fast-growing softwoods are farmed, with harvesting occurring in rotation.

Producing composite wood requires binders to fuse wood particles and fibers together. Urea–formaldehyde and phenol–formaldehyde binders are commonly used, but these toxic chemicals cause poor indoor air quality as a result of their carcinogenic compounds. Specifying products using alternative binder compounds like polyvinyl acetate or soy are safest. LEED points are earned for using nonformaldehyde-based cabinets, countertops, moldings, and trims.

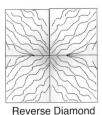

Reverse Diamond   Diamond   Sunburst

Book match   Box match

FIGURE 3.34 Pattern matching of veneers yields a variety of designs that emphasize the grain of the wood.

FIGURE 3.35 The interior of this Victorian–era home features classic millwork details as seen in the crown molding, window trim, wall paneling, and mantelpiece.

comprise the majority of trims and moldings installed in new construction projects today. Stock moldings are sold in paint-grade or stain-grade woods, come in a wide range of profiles, and are relatively inexpensive. Paint-grade wood is the least expensive and is made from pine or other softwoods in which the grain pattern is not needed. Stain-grade woods, like oak, cherry, walnut, or mahogany, feature prominent grain patterns enhanced by staining.

Molding profiles are based on designs first used by the ancient Greeks and Romans on temples and civic buildings (Figure 3.36). The profile is the shape of the molding as viewed from the side. Profiles are grouped together for visual effect; some are complex in design whereas others are rather simple. Profile groupings are chosen to coordinate with the overall style of the home. Unadorned, flat moldings and trims with small quarter-round fillets are well suited to the simple purity of arts and crafts-style homes. Anything more ornate would be out of character.

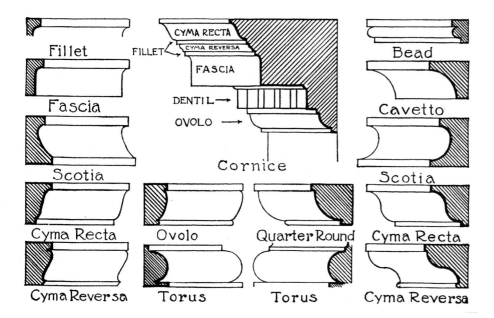

FIGURE 3.36 Molding profiles used by ancient Greek and Roman architecture are shown in this drawing.

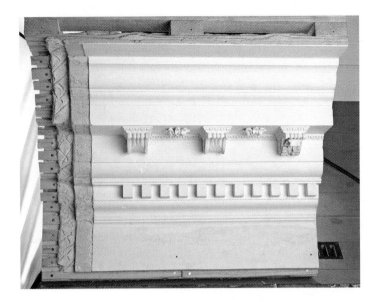

FIGURE 3.37 Prefabricated crown moldings made from plastics, resins, and Styrofoam are easy to install and can be painted any color.

Elaborately designed cornice moldings made from cast polystyrene or plastic are available as inexpensive substitutes for milled wood and, when installed high above discerning eyes, may be passed off as painted wood (Figure 3.37). Some purists may debate this observation; however, prefabricated moldings can achieve the desired effect without extending the construction budget. Prefabricated moldings made from these materials are preprimed for painting.

Baseboards are installed where the floor meets the wall to protect it from the nicks and bumps of vacuum cleaners, furniture, and shoes. Baseboards hide any expansion gaps left between the flooring material and the wall, and are sized proportionately to the ceiling height of the room. Eight-foot ceilings usually have 4-inch baseboards whereas 6-inch baseboards might be used in rooms with 10-foot ceilings. Often, baseboards and the trim around doors and windows have the same profile; they may be unadorned or highly expressive, designed to blend in with the overall aesthetic of the home (Figure 3.38). Trim around doors and windows is a necessary architectural element to cover gaps left between the wall and door or window assembly. The gaps are filled in with weather sealant, and then trim is applied around the openings on both the interior and exterior walls.

MILLWORK FOR KITCHENS AND BATHROOMS

Millwork for kitchens and bathrooms include the cabinets, vanities, and cupboards that go into making these rooms functional. Stock cabinetry is available in a wide range of styles and is priced from low to high according to the materials, finishes, and details like raised panel fronts or glass-fitted frames (Figure 3.39). Manufactured in standardized component sizes, stock cabinetry allows for easy customization of kitchen layouts.

Base cabinets are built 24 inches deep and 34½ inches high, with widths ranging from 9 inches (for things like vertical tray storage) to 42 inches wide for fitting oversize farm-style sinks. Upper cabinets are 12 inches deep and vary in height from 12 to 40 inches. The same widths for base cabinets are available for upper cabinets to ensure a balanced layout. Kitchen cabinet options include conveniences like pull-out bins for trash and recycling, and slide-out drawers for easy access. Custom cabinets take longer to manufacture, and significant lead time is required

FIGURE 3.38 Window trim and cornice molding give an authentic touch to this new arts and crafts-style kitchen.

to build them. They are specially made through a cabinet supplier and cost more money than *stock cabinets;* however, bench-built cabinetry, made by a finish carpenter, is the most expensive.

After the base cabinets are installed, a template is made for the finished countertop. The template follows the configuration of the base cabinets, marking locations for sinks and faucets where holes will need to be cut from the countertop material. The countertop material is then fabricated based on the template. Countertop materials like stone, concrete, plastic laminate, stainless steel, tile, or wood are specified for performance, and aesthetics. Natural stone like granite or soapstone is durable, holds up to heat, and requires little maintenance other than an annual application of sealer. Engineered stone products like brand name Silestone Quartz offer the durability of natural stone without the maintenance. Solid surfacing materials like brand name Corian are made from molded synthetics and are microbial. Natural stone, ceramic tile, and wood countertops are not microbial, so informing clients about the maintenance and hygienic properties of each countertop material is important for specifying the right material for the job.

FIGURE 3.39 Upper cabinets feature glass panel fronts, bead board base cabinets, and a blending of maple and white quartz countertops.

Plumbing Fixtures and Faucets

Selecting faucets and plumbing fixtures is often an overwhelming task for many clients. A wide range of products is on the market and there are many decisions to be made. Does the client want low-profile toilets with silent flushing systems? What about having a kitchen faucet with a pull-out sprayer? Or a hot-water dispenser on the sink? Perhaps the clients want a recessed floor sink in the mudroom, or a sink that uses sonic waves to wash their delicates (Figure 3.40). Plumbing needs must be addressed early during the planning stages to ensure clients get what they want before walls are enclosed and finished.

FIGURE 3.40 This combination laundry and mudroom features special-purpose sinks, like a floor sink for cleaning muddy shoes.

FIGURE 3.41 Wall-mounted faucets are a practical choice for this glass wall-mounted sink.

As stated in Chapter 2, plumbing fixtures and faucets must be specified before the plumbing lines are finalized. Valve sizes and fittings range in size from one-half inch to three-quarters of an inch and must be coordinated with faucets, shower heads, and tub fillers to ensure the components will fit together. In addition, the plumber must know how the faucets will be mounted for each fixture—whether they should be mounted on the wall, on the countertop, or the deck edge of the tub or sink (Figure 3.41).

When selecting faucets, clients usually emphasize the desired style and finish, although water-saving features should not be overlooked. In kitchens, faucets are deck mounted, wall mounted, or countertop mounted, depending on the type of sink. Sinks with predrilled holes must be ordered for deck-mounted faucets, and the configurations must match those of the faucet. Faucets with pull-out sprayers are available for easy washing, built-in dispensers keep soap nearby for hand washing, hot-water dispensers are convenient for sanitizing dishes, and pot fillers over stoves make cooking easier (Figure 3.42).

Bathroom faucets include those specified for sinks, showers, bidets, and tubs, and the style and finish match toilet flushers, towel bars, toilet roll holders, and soap dishes (Figure 3.43). Manufacturers of kitchen and bathroom faucets and fixtures make it easy by offering coordinated sets for all these needs. The plumber or plumbing supply representative should review the order to ensure all not-so-obvious parts, including drain covers and connectors, are specified.

Although the majority of sinks sold in America are made from porcelain, metal sinks, glass sinks, and sinks made from natural stone or composite resins find their way into kitchens and bathrooms. Cast iron sinks coated with enamel keep water hot for longer periods of time than those made from other materials. Sinks made from synthetic resins and polymers are easily molded into any shape or configuration and are sold as all-in-one units, including the countertop.

Before specifying sinks, it is important to know whether the sink will be undermounted or dropped into a counter or vanity top. This will depend on the type of countertop material. For example, if the countertop is a solid surface, like stone, wood, or synthetic resin, sinks should be undermounted (Figure 3.44). Undermounted sinks make it easier to scrape food waste from the countertop right into the garbage disposal. However, countertop materials like tile or plastic laminate

GO GREEN Water-Saving Devices

Standing under the shower for a good long soak is no longer politically correct these days because our water resources are in danger of pollution and depletion. Installing an aerator on faucets saves water by injecting air into the stream of flowing water, thereby reducing the amount of water needed to wash hands or take a shower. In addition, manufacturers have developed low-flow faucets for sinks, tubs, and shower heads that further reduce the amount of water needed for washing. Moreover, many municipal codes require that only *water closets* using a maximum of 1.6 gallons are allowed to be installed in new residential construction.

FIGURE 3.42 A pot filler offers maximum convenience for any cook in the kitchen.

require self-rimming sinks that seal against water damage to the OSB or plywood substrate (Figure 3.45).

Residential toilets are made from *vitreous china* that is impervious to water and strong enough to resist cracking or chipping. When a toilet is flushed, water in the bowl is forced down the waste stack, and a float inside the tank rises up as new water is piped in to refill the bowl. When the float reaches a certain level in the tank, it stops the new supply of water from coming in so the toilet does not overflow. As a conservation measure, efficiency codes in most states require toilets to use less than 1.6 gallons with each flush.

Used in Europe for decades, a bidet is a basin similar in design to a toilet, but without a seat, and is fitted with faucets that dispense hot and cold water for personal cleansing (Figure 3.46). A bidet has no flushing device; water drains the same as a sink. Bidets have hit the American market in greater numbers than the previous decade, accompanying toilets in high-end master bedrooms.

FIGURE 3.43 A fully adjustable sliding shower head allows for easy use by adults or children alike.

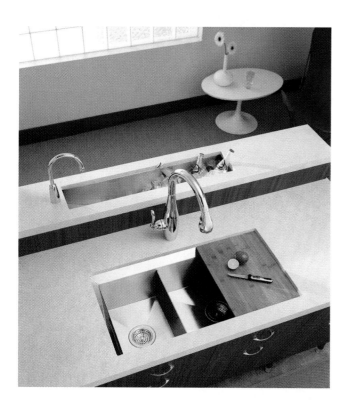

FIGURE 3.44 Seamless countertops like those made from solid surfacing materials work best with undermounted sinks for easy cleaning. Deck-mounted goose-neck faucets provide clearance for rinsing tall bottles or pots.

FIGURE 3.45 A self-rimming specialty sink is dropped into a prepared countertop with a tiled backsplash.

FIGURE 3.46 A bidet offers practical solutions to hygienic cleansing.

FIGURE 3.47 Framing for a spa tub must incorporate an access panel for servicing the whirlpool equipment.

FIGURE 3.48 A steam shower featuring massaging shower jets is enclosed in frameless, sealed glass doors.

Bathtubs are either built into a wall on three sides or are freestanding, and are available in vitreous china, enameled cast iron, *fiberglass*, or metal, like copper or stainless steel. Tub sizes range from a standard 3 by 5 feet upward to large whirlpool types. The mechanical systems running the motor and jets in a whirlpool tub must be accessible in case repairs are needed, and the surrounding panels enclosing a whirlpool tub need to be removable (Figure 3.47). Whirlpool tubs and large soaking tubs require additional reinforcement in the floor joists and subfloor, with the ability to support 70 gallons or more of water. These large tubs are not designed to conserve water, and a code check with local municipalities must be done before specifying these products.

Stand-alone showers, those not included in the tub enclosure, are site built with concrete shower pans sloped toward the drain and poured onsite to the desired size and shape. The concrete pan is finished with ceramic tile or stone. For conventional square or rectangular shapes, a shower pan made from formed metal or fiberglass can be installed and needs no floor finish. Shower stalls and tub enclosures are finished with ceramic tile or stone over waterproof cement board (Figure 3.48). Remaining walls in these humid bathroom spaces should be enclosed with water resistive gypsum wallboard, often called green board because the paper covering the core is green.

Informative Websites

Environmental Bamboo Foundation: **www.bamboocentral.org**

Green Seal: **www.greenseal.org**

Gypsum Association: **www.gypsum.org**

Northwest EcoBuilding Guild: **www.ecobuilding.org**

National Council for Interior Design Qualification: **www.ncidq.org**

Commercial Construction and Systems

4

Commercial Construction
Building the Envelope

Building Basics

Major cities around the world are recognized by their landmark skylines with a host of unusually shaped *high-rise* towers that appear to defy gravity as they reach skyward. The earliest beginnings of skyscrapers date back to 1890 with the construction of the 10-story Wainwright Building in St. Louis. The tallest building for the time period set competition in motion to see who could build a taller building. In their quest to build taller structures, architects and engineers developed construction techniques using new materials like steel and reinforced *concrete*. Like all construction projects, the process of building high-rise towers starts with the foundation, then advances to the structural framework, and stops at the roof. Nevertheless, not all commercial buildings are sleek high-rises; there are *low-rises, mid-rises,* strip malls, and small freestanding structures as well.

Building codes for commercial buildings are stricter than codes for residential occupancies because the risk is greater. Often, these buildings are occupied by a large concentration of people unfamiliar with the building layout or location

The skyline of Chicago features modern high-rises made from steel and concrete clad with shimmering glass.

of exits. For example, movie theaters accommodate many people within a dark auditorium in fixed seats. Safe evacuation in case of fire or smoke is a priority for the design of these spaces, and codes address these specific *egress* issues. Moreover, office buildings feature floor after floor of workers, but visitors may be unfamiliar with the locations of exits or fire stairs within the building.

First and foremost, *prescriptive codes* and *performance codes* mandate that commercial buildings must be designed and built in ways that allow occupants to get out of the building quickly and safely in case of an emergency. Furthermore, codes are in place to protect the occupants of a building from being exposed to hazardous materials or toxic indoor air quality, and universal access codes require that the building be fully accessible by persons with physical challenges. Initial code research and following the codes throughout the planning and construction phases of a project protect occupants and reduce the risk of injury or death. However, codes provide only minimum guidelines. Prudent engineers, architects, and designers should take care to provide maximum safety for all building occupants. In addition to these code requirements, *occupancy classifications* or *user groups* identify how hazardous the operations are for the building occupants, which in turn determine which codes must be followed.

FOUNDATION SYSTEMS

The first stage of commercial construction is preparing the site for a foundation system. For high-rises, this starts by excavating the building site. The higher the structure, the deeper the excavation, sometimes extending several feet below ground to reach bedrock (Figure 4.1). The type of foundation system used is determined by the live and *dead loads* (or overall weight) the skeletal framework will carry after the building is completed. Dead loads include the weight of the walls, floors, roofs, and ceilings; *live loads* are the weight of interior furnishings, equipment, and people inside.

Foundation systems for multistory commercial buildings include poured concrete *caissons,* steel piles, and concrete or steel columns and *footings* (Figure 4.2). The type of foundation system used depends on the stability of the soil and how deep into the ground the foundation must go to reach solid rock. Concrete caissons

FIGURE 4.1 A steam shovel is used to excavate the site in preparation for the foundations of a new office tower in New York City.

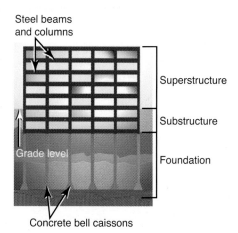

Steel beams and columns

Superstructure

Substructure

Grade level

Foundation

Concrete bell caissons

FIGURE 4.2 This drawing illustrates the key sections of a high-rise building: the foundation, substructure, and superstructure.

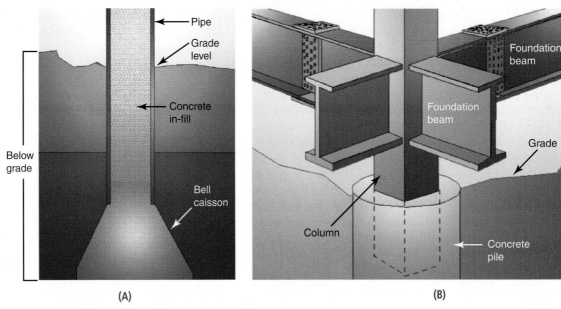

FIGURE 4.3 (A) A caisson is shaped with a flange, or bell, at the bottom to provide a stable footing. Drilled augers dig deep into the ground to extract soil, which is then filled with reinforced concrete. (B) The structure's steel framing is then anchored to these caissons.

are set with equipment similar to what oil companies use to drill into the ground. Long, narrow channels are excavated and then the hole is filled with reinforced concrete (Figure 4.3). The building's structural steel framework is anchored to these concrete caissons to create a sturdy support.

Another type of deep foundation system utilizes pile drivers to hammer steel *pilings* into the ground until they reach stable soil or rock (Figures 4.4 and 4.5). The framework of the structure is then attached to and supported by these pilings. Newer to the market, helix piles with augerlike blades are faster to install, because the blades rotate as they are pushed downward, cutting through layers of soil (Figure 4.6).

FIGURE 4.4 A pile driver hammers a steel post into the ground, which will serve as a foundation pier for a new building.

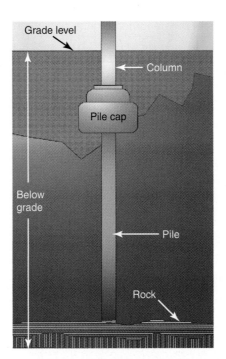

FIGURE 4.5 After foundation piles are driven down into firm rock, construction on the sub- and superstructures can begin.

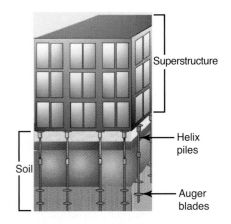

FIGURE 4.6 This drawing illustrates a foundation system comprised of helix piles for a mid-rise building.

FIGURE 4.7 Concrete is poured onsite over a framework of rebar for a slab-on-grade foundation.

FIGURE 4.8 Formwork for poured concrete piers is set up on concrete footings.

Slab-on-grade foundations are commonly used for building strip malls, grocery stores, and small office buildings where soil and climate conditions allow (Figure 4.7). When a foundation must extend beneath the frost line or below grade to reach stable soil, concrete footings may be poured in place on prepared soil. Concrete footings reinforced with steel rebar are sized according to the load they will carry based on their *compressive strength* (measured in *pounds per square inch*). After the footings are set, the *formwork* is erected for *site-cast concrete* piers to complete the foundation system (Figure 4.8).

STEEL FRAME STRUCTURES

Steel frame buildings are constructed with a structural framework of *prefabricated* beams, columns, and trusses calculated to support the loads of the finished building. These architectural elements are sized by a structural engineer according to the *tensile strength* of the steel. Steel beams are made at a steel mill to the exact specifications, then are shipped to the building site where they are joined together by a crew of welders and construction workers. A prewelded clip angle along with rivets and connector panels are used to connect beams to columns to ensure a stable bond (Figure 4.9). Prefabrication allows for quick construction when the materials arrive onsite (Figure 4.10).

FIGURE 4.9 A worker uses an impact wrench to tighten the bolts on a clip angle attached to a steel girder.

FIGURE 4.10 A crane hoists a steel girder to construction workers who will then bolt it into place.

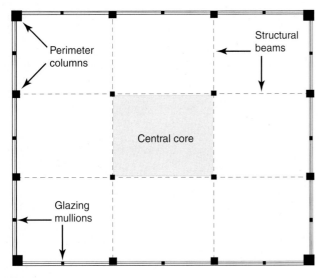

FIGURE 4.11 This floor plan for an office building shows the placement of perimeter and interior columns that support the structure.

A structural engineer uses charts to size the columns, beams, and girders for high-rise buildings according to the loads they will carry. Columns and beams on the lower floors are larger than those nearest the top. As the *superstructure* compresses downward from the floors above, the total weight is distributed onto foundation piers. Interior load-bearing columns are spaced on a grid, with bay widths typically running every 15 or 20 feet (Figure 4.11).

Steel is made from an alloy of iron and less than 2% carbon. The carbon gives iron more tensile strength, allowing it to "bend" under weight. Steel has a melting point of 2,600 to 2,800°F, and structural beams and columns are made through an extrusion process. The shape of these *extruded steel* elements forms the letter "I" and are commonly called *I beams*. Fire codes require all structural steel components for commercial buildings be treated with a *fire resistive* material. Designed to guard against structural failure in case of fire, spray-on fireproofing, or encasing the columns and beams with concrete, or gypsum wallboard, will meet this code requirement (Figure 4.12).

As the steel framework is constructed, added reinforcements like diagonal bracing are incorporated to stabilize the skeletal framework against *lateral forces* like strong winds or seismic changes (Figures 4.13 and 4.14). Although the steel frame requires some form of additional reinforcement, high-rise office towers are designed to bend with the wind, up to 18 inches off the vertical axis. The building's sway is instrumental in keeping the structural framework safe, although some office workers complain of motion sickness on extremely windy days.

Other methods of stabilizing the steel framework include anchoring the steel trusses, *metal decking*, and poured concrete floor system to the rigid vertical core of the building. The concrete *central core* inside a building is a vertical shaft where plumbing lines, electrical feeds, elevator banks, and stairwells are housed (Figures 4.15 and 4.16). Rigid concrete sheer panels attached to the exoskeleton is another method of stabilizing steel frame structures.

LEED The Way

Take the Gold!

Hearst Tower in New York City became the first office building in the state to receive Gold Certification from the U.S. Green Building Council for its achievements for sustainability. Designed by Norman Foster and completed in 2006, the office tower's unusual triangular-shaped structural system uses 20% less steel, with 89% of it made from recycled steel. The building also consumes 25% less energy than ordinary office buildings because of its low E-coated exterior glass cladding, and a ventilation system that uses outside air for 75% of the year. Other features include motion sensor lighting, low-VOC interior finish materials and furnishings, and a water collection system on the roof, which provides gray water for air-conditioning systems and landscape irrigation.

FIGURE 4.12 Spray-on fireproofing is visible on the steel beams and girders in this new building.

FIGURE 4.13 A low-angle view of the Bank of China in Hong Kong shows the diagonal bracing, running the full height of the building, used to stabilize the steel frame.

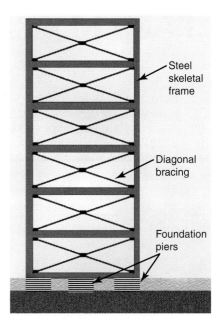

FIGURE 4.14 This diagram shows a system of diagonal bracing for steel frame construction. Diagonal bracing reduces lateral loads on the structure like those caused by earthquakes.

Because building loads are supported by columns, beams, and girders, and are distributed to the foundation system, the enclosing weatherproofing for the interior (essentially, the exterior cladding material) is nonload bearing. Exterior cladding materials like tinted and reflective glass, noncorrosive metal, or stone are chosen for their aesthetics as well as their ability to reduce heat gain or loss within the building envelope (Figures 4.17 and 4.18). This nonload-bearing cladding system called a *curtain wall* is installed over the steel frame. The curtain wall system is supported by a framework designed to carry the weight of the enclosing material, resist wind loads, and tolerate a degree of seismic changes.

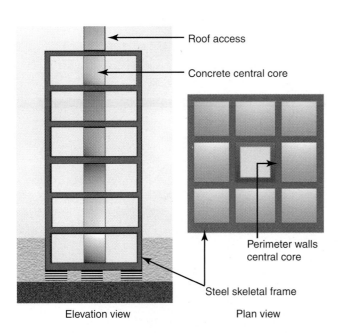

FIGURE 4.15 Buildings with a central concrete core designed for locating elevators and stairwells provide additional reinforcement to steel frame construction.

FIGURE 4.16 The concrete central core is the first structure built on this construction site for a multistory office building. The core will house fire exit stairwells, mechanical ductwork, and the elevator shaft.

LEARN More

Skin and Bones Architecture

German architect Ludwig Mies van der Rohe (1886–1969) contributed to the development of the modern glass skyscraper with a model he designed in 1919. Before becoming the director of the Bauhaus in 1930, his innovative architectural achievements included all-glass houses and the design of a model of an all-glass 30-story building. The model of a glass tower purported Mies's "skin and bones" approach to high-rise construction; a steel skeletal frame (the bones) carried the load of the structure, making it possible for nonload bearing curtain walls of glass (the skin) to enclose the building. The structure proposed in his model proved too advanced for construction techniques at the time, and Mies's concept for a steel and glass tower was not realized until the 1950s. Limited construction technology coupled with the lack of sophisticated HVAC systems kept the concept of all-glass high-rises on the drawing table until improvements in tinted plate glass and HVAC systems were realized. Mies's model from 1919 became a prototype for glass box-style modernism throughout the 1950s and 1960s, and subsequent late-modern styles of the 1970s.

A model of the skyscraper envisioned by Ludwig Mies van der Rohe in 1919.

CONCRETE STRUCTURES

Throughout the world, concrete is the most common material utilized in commercial building construction because, unlike steel, the raw materials for making concrete like aggregate, lime, water, and cement, are readily available. Concrete building

FIGURE 4.17 The exterior cladding on this new office building is made from laminated glass. Although the glass itself is nonload bearing, it is designed to resist forces from high winds and it provides acoustical and thermal protection.

FIGURE 4.18 Lightweight aluminum panels enclose this high-rise in Tokyo.

LEARN More

Tall, Taller, and Tallest

A host of major cities around the globe—from New York to Taipei—at one time laid claim for having the tallest building in the world, but it was the opening of the Empire State Building in 1931 (the tallest building in the world at that time) that began the race to see who could break its record by building higher than its 102 stories. The term *skyscraper* was first used in 1890, when the Wainwright Building in St. Louis, designed by Louis Sullivan (1856–1924), was completed. This office building dazzled the public with its 10-story height—a building so tall it "scraped the sky." Sullivan designed his building with steel framing, which made it possible to build a building higher than anyone had attempted before.

The Empire State Building rises high above all the other nearby buildings.

Soon, improvements in building technology like elevators and air-conditioning enabled architects to design taller structures without sacrificing convenience and comfort. Forty years after the Wainwright Building opened, in 1931, New York City possessed the tallest building in the world. The 1,453-foot-high Empire State Building boasted 1,250 feet of structure topped with a massive spire and antennae. This impressive landmark was clad in limestone with small, operable windows (there was no air-conditioning at the time) and the fastest elevators available at that time.

For the next 40 years, the Empire State Building remained the tallest building in the world until the World Trade Center towers were built in 1972. The now-infamous Twin Towers in New York City reached 110 stories into the sky and held the world's record until the building of the Sears Tower (now referred to as *Willis Tower*) in Chicago in 1974: completed at a height of 1,450 feet of structure to the roof, and 1,730 feet including antennae. As of this writing, the tallest building in the world is the Burj Khalifa, which rises to a height of more than 2,600 feet with 167 stories. The structure is supported by a framework of prestressed concrete. Building the tower led to advancements in concrete-pumping technology to reach the top floors.

Burj Khalifa, the world's tallest skyscraper as of this writing, incorporates a concrete structural frame that towers to a height of more than 2,600 feet.

materials are either prefabricated or poured onsite. Concrete is poured into formwork or molds, giving the structural element its shape. Chemicals in the concrete compound heat up and the curing process begins. It is important to note that concrete loses its strength if it sets or cures too quickly. When the concrete has fully cured, it has a rocklike hardness and strong compressive strength capable of supporting direct loads (Figure 4.19).

Site-cast or prefabricated concrete beams and columns are sized according to the weight they will support. Site-cast concrete utilizes a system of wooden or metal forms to set the desired shape, size, and thickness for the finished building components like foundations, walls, floors, and ceiling panels. A network of steel rebar, adding strength and dimensional stability, is inserted into the formwork before the concrete is poured (Figure 4.20). After the concrete sets, the formwork is removed and the rough surface texture of the concrete is either left as is or is *sandblasted* to remove signs of the formwork.

FIGURE 4.19 City Spire in New York was the second largest reinforced concrete building in the world when it was built in 1987 by Helmut Jahn.

Earthquakes

Without warning, a natural disaster such as an earthquake can destroy buildings in their wake unless precautions are taken during the design and construction of the building. In areas prone to severe earthquakes, such as southern California, building codes have been strengthened in efforts to avoid serious injury caused from structural failure. After the 1994 earthquake in Northridge, a suburb of Los Angeles, current building codes for new construction as well as codes for retrofitting older buildings were improved to strengthen the rigidity of a building. Seismic changes occurring during earthquakes put lateral stress on the structural system of the building. Earthquake codes in California now require the building *substructure* and superstructure to be strengthened with rigid shear walls for resisting lateral forces, the columns must be made stronger than normally necessary for nonearthquake areas, and beam and column connectors must be designed to flex without breaking to help stabilize the building against seismic forces. Moreover, it isn't enough to take preventive measures on the structural part of the building alone; codes are set in place to guard against nonstructural elements like water pipes, heating and air-conditioning ducts, and ceilings fixtures from falling during an earthquake and causing injury.

A buckled and cracked cement column on the corner of a high-rise building resulted from the earthquake that hit Northridge in 1994.

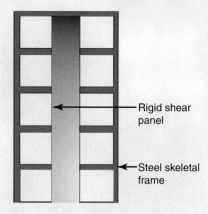

Rigid shear panel

Steel skeletal frame

Rigid shear panels added to the exoskeleton of a steel frame building aids in resisting lateral forces like those caused by earthquakes.

FIGURE 4.20 Formwork with rebar inserts is ready for pouring concrete to make supporting columns for a new building.

Prestressing concrete gives beams and columns the compressive strength of concrete with the tensile strength of steel. This is achieved by pouring wet concrete into formwork that contains steel reinforcements that are placed under tension. During the concrete curing process, the tension is released from the steel reinforcements to give the concrete beams and columns maximum strength and flexibility. Prestressed beams allow for longer spans between columns, leaving the interior with fewer obstructions (Figure 4.21). Building with prefabricated concrete materials speeds up the construction process because there is no wasted time waiting for the concrete to cure. Prefabricated wall panels are shipped to the site and erected onto a prepared foundation with cranes. Just like steel frame construction, the exterior cladding on concrete structures is nonload bearing (Figure 4.22).

LEARN More

Brutalism

A new aesthetic—a style called *brutalism*—emerged during the mid 20th century as poured concrete buildings appeared in urban projects across America.

Exterior view of the Art and Architecture Building by Paul Rudolph, 1959–1963, New Haven, Connecticut.

Architects working in this new style such as I. M. Pei, Paul Rudolph, Louis Kahn, and Philip Johnson wanted the impressions left over from formwork molds like nuts and bolt marks, plywood grain patterns, and joint seams to remain visible on the finished building. Instead of sandblasting the cured concrete to a smooth finish, these imperfections became part of the character of the building exterior. Reyner Baham's *The New Brutalism* (1966) perpetuated the style, but by the 1980s the style was no longer popular

Example of surface texture provided by inserting a plastic liner inside the formwork for pouring concrete.

Tilt-wall concrete construction is a quick and cost-effective method for building low-rise shopping malls, big-box stores, and industrial buildings. Tilt-wall construction is either site cast or prefabricated. Concrete wall panels are cast into forms laid on the ground at the building site. After the concrete has cured, they are hoisted into place. Locations for doors and windows in tilt-wall concrete construction must be formed at the time of the pour, regardless of whether they are site cast or prefabricated (Figure 4.23). These unfinished concrete walls are finished with a variety of aesthetic treatments ranging from stucco, paint, metal cladding, or brick veneer (Figure 4.24).

FIGURE 4.21 An interior view of an office building under construction shows large concrete columns supporting precast concrete ceiling beams.

FIGURE 4.22 Precast concrete structural members were used to build this low-rise office building.

FIGURE 4.23 Precast panels are temporarily supported by steel braces in this tilt-wall construction project.

FIGURE 4.24 A new big-box store built with tilt-wall construction features a range of decorative details on its concrete exterior.

FIGURE 4.25 A worker sits on open-web steel joists as he prepares to attach metal decking for the flooring system in this new building.

Floor Systems

Floor systems for commercial buildings provide support for the floor above and form the ceiling for the floor below, and are constructed with either steel beams or *open-web steel joists* (Figure 4.25). Open-web steel joists are prefabricated into segment lengths, widths, and depths sized according to the loads they will carry. These structural elements are then shipped to the site and are connected to the steel girders or beams on a steel frame building, or are attached to load-bearing masonry walls (Figure 4.26). Metal decking is then secured over steel beams or joists to provide a continuous platform onto which lightweight concrete is then poured (Figure 4.27). Alternatively, precast concrete panels with interlocking joints are also used, eliminating the need for metal decking.

FIGURE 4.26 Open-web steel joists are bolted to masonry walls to provide the structural framework for a flat roof.

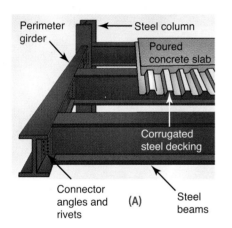

(A)

Perimeter girder · Steel column · Poured concrete slab · Corrugated steel decking · Connector angles and rivets · Steel beams

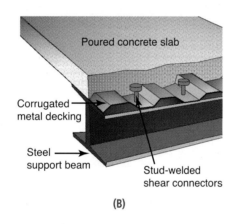

(B)

Poured concrete slab · Corrugated metal decking · Steel support beam · Stud-welded shear connectors

(C)

FIGURE 4.27 (A) These drawings illustrate how corrugated metal decking is laid and connected over floor joists. (B) The decking provides a sturdy substrate for pouring a concrete floor. (C) A view looking up to an open ceiling reveals the underside of metal decking supported by open-web steel trusses.

Roof Systems

Commercial buildings designed with flat roofs have a few advantages over those with peaks or slopes. A flat roof provides plenty of space for housing HVAC equipment and is constructed the same way as the floor system (Figure 4.28). The weight of the equipment installed on the roof is taken into consideration when the structural engineer sizes the steel trusses and calculates the thickness for the concrete roof slab. In addition, access to the roof must be available from inside the building for equipment maintenance and repair. In low-rise buildings, nonstructural *parapets* are designed to hide from sight the rooftop equipment from the street below.

Articulated roof designs are formed with steel trusses prefabricated into segment lengths and dimensions that will support the finishing material, whether glass panels for skylights or precast concrete panels (Figure 4.29). Regardless of whether the roof is pitched, sloped, or flat, the roofing material must be watertight and be able to withstand high wind forces. Rubber membrane covered with gravel is most commonly used as a roofing material for commercial buildings,

FIGURE 4.28 Mechanical equipment is seen on the rooftops of high-rise buildings in this aerial view of downtown Philadelphia.

FIGURE 4.29 Skylights are a significant feature of this downtown skyline.

The roof of the Intercultural Center at Georgetown University is covered with solar panels that have remained in operation since 1984. The project was developed between Georgetown University and the U.S. Department of Energy in response to the first energy crisis during the 1970s. The angled roof faces south, allowing the photovoltaic cells to capture energy from the sun, which is then converted into electricity.

Unfortunately, the cost-effectiveness of equipment installation related to energy dollar savings has not yet been bridged, and the advancement of using solar energy in commercial structures is still the exception instead of the norm. Despite slow progress toward viable solar energy usage, after Georgetown's inaugural launch other universities were challenged to incorporate measures for increased energy efficiency on their campuses. To see what measures your campus has adopted toward saving the planet, visit the website for the Association for the Advancement of Sustainability in Higher Education at www.AASHE.org.

Photovoltaic modules cover the roof of Georgetown University's Intercultural Center in Washington, DC, providing power to the building and campus.

although advances in roofing technologies support green design by offering a wide range of materials, including photovoltaic shingles and vegetative roof covers. For more information about grass roofs for commercial buildings, visit the website of Whole Building Design Guide from the National Institute of Building Sciences (www.wbdg.org/resources/greenroofs.php).

Doors and Windows

Commonplace in commercial buildings are ground-level facades with large expanses of glass fitted with glass doors that give views of what lies inside—whether a large lobby, a reception desk, or a waiting area (Figure 4.30). The doors and fixed glass panels are made from safety glass and are secured in structurally stable metal frames designed to stand up to frequent use by the constant flow of people. Exterior doors for commercial buildings must comply with codes that ensure safety and accessibility for the occupants of a building, in addition to providing security, privacy, thermal protection, and acoustic control. They must be durable enough to withstand the comings and goings of people either casually or in panic-induced situations (like escape from fire or smoke).

Exterior doors leading to service areas like parking garages, loading docks, or public ways are made from materials based on the occupancy hazard. For example, doors to parking garages are usually made from fire-rated metal and glass. Glass viewing panels ensure visibility between persons entering or exiting, and the metal door keeps carbon monoxide from entering into the building from the garage. Moreover, doors to parking garages

FIGURE 4.30 The entry into this low-rise building features large expanses of fixed glass with metal framed glass doors.

LEARN More

Occupancy Classifications and Exits

Building codes for commercial structures are stricter than those required for one- or two-family dwellings and require careful research before beginning a project. Unlike residences, commercial buildings are used by the general public, where concentrations of people gather in conditions that are unfamiliar to them. Being unfamiliar with the layout of a building and how to get to an exit could become hazardous in case of emergency evacuation.

Knowing the occupancy classification of the proposed building is the starting point in researching the codes it requires. The occupancy classification assesses the hazard of the business within the building and it is used to calculate how many exits, or means of egress, are needed. For example, a multiroom movie theater has limited passageways and aisles for people to get in to or out of rows of fixed seats. In the case of an emergency, during which people will need to exit in a hurry, these fixed aisles could become hazardous in a panic situation. Because of the high concentration of people in a confined space like a movie theater, the project carries an "Assembly" occupancy classification, and building codes require that these spaces have an adequate number of exits that lead directly to the outside or to a path of egress.

For other occupancy classifications, the number of exits required in a building is determined by the *occupancy load*—a factor indicated in the codes and calculated according to the square footage of the space. In "Business" classifications like general office buildings, an occupancy load for each floor of a building of 500 persons or less requires two exits or means of egress. If the occupancy load is more than 500 persons but less than 1,000 persons per floor, three exits are required. For those occupancy loads of more than 1,000 persons, four exits are required. Occupancy classifications identify the user group of a building and are used to determine which codes must be followed:

- **Assembly Group A:** The use of a building for gathering a concentrated amount of people in a room or space like theaters, restaurants, art galleries, museums, and stadiums

- **Business Group B:** The use of a building for offices, educational facilities above the 12th grade, animal hospitals, and banks

- **Educational Group E:** The use of a building for schools for children through the 12th grade, including daycare centers with more than five children older than 30 months of age

- **Factory Group F:** The use of a building for industrial purposes, including those used for manufacturing or product assembly

- **Hazardous Group H:** The use of a building for hazardous operations such as chemical laboratories, and power plants

- **Institutional Group I:** The use of a building by persons requiring custodial care, including hospitals, nursing homes, and prisons or jails

- **Mercantile Group M:** The use of a building by persons distributing merchandise or goods to the general public, including department stores, vehicle service stations, and grocery stores

- **Residential Group R1, 2, 4:** The use of a building for commercial residency occupancy, including hotels, apartment buildings, and dormitories

- **Storage Group S:** The use of a building for storage of merchandise including food, furniture, appliances, and so on

- **Utility and Miscellaneous Group U:** The use of a building for ancillary operations or services such as aircraft hangers, greenhouses, sheds, and stables

For more information and complete descriptions of occupancy classifications and their subgroups, refer to the International Code Council (www.icc.org).

are equipped with automatic door closers to ensure the door will remain closed. A code check done before specifying exterior doors ensures that size, materials, and locations meet building and energy codes, and universal access codes.

According to the Center for Sustainable Building Research, studies in human factors have established that workers are more productive when exposed to natural lighting. This information endorses why executives occupy corner offices as a measure of status; corner offices have two walls of windows compared with other offices sandwiched along the building perimeter. Windows in commercial buildings are either fixed glass panels or fully operational. Fixed glass windows in high-rise office towers are incorporated into the curtain wall system of the building. For operable windows in smaller, stand-alone commercial buildings, large expanses of glass or wide ribbon windows run from one end of the building to the other (Figure 4.31). Designers must solve both the aesthetic appeal and human behavioral needs when specifying windows for commercial spaces.

CAUTION

Exit Doors

International Building Codes and the Americans with Disabilities Act provide detailed requirements for the exit doors in commercial buildings. Exit doors must be hinged and of the swinging type, and must have a minimum clear

Revolving doors like these in the Grand Hyatt Hotel in Beijing are used in buildings to curtail thermal impact loads on HVAC systems by frequent opening and closing of doors. However, revolving doors are not to be relied on for emergency exits and are flanked on either side by fully operable swinging doors.

width opening of 32 inches when the door is fully opened. The height of all exit doors must reach 80 inches; if automatic closers are installed, a clear height of 78 inches must be maintained. Door assemblies must be equipped with easy-release hardware activated by asserting light pressure to the handle, bar, or plate, and the doors must swing open into the direction of travel.

The codes may allow exceptions for specific occupancy classifications and/or door types. For example, revolving doors are allowed as exits in certain occupancies as long as they are not more than 50% of the total required number of exits. Furthermore, revolving doors must have a regular, swinging-type door within 10 feet for an alternate means of egress.

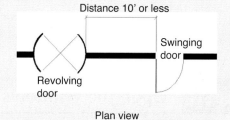

This architectural plan symbol represents a revolving door.

FIGURE 4.31 A view through the floor-to-ceiling glass wall in this conference room reveals traditional sash-style windows in this newly converted warehouse.

To meet the exact dimensions of these variable openings, commercial windows are made to order. Commercial windows are specified for their performance, ease of maintenance, energy efficiency, and cost. Expansive openings are controlled with exterior sun shields or deep roof overhangs to eliminate direct sunlight from reaching the glass, or with interior shades and blinds.

Unless treated, large expanses of glass are the main source of heat gain or heat loss in a building and must be controlled to comply with energy codes. Thermal control for windows is accomplished through a variety of methods, including the application of a UV film, using *laminated glass* filled with an inert gas, or by sandwiching multiple layers of glass inside the window or curtain wall assembly.

Glass treated with *low E-coatings* and rated for high solar gain is specified for cold-weather climates. These windows maximize the transfer of heat from the sun, through the glass, and into the interior. For warm weather climates, commercial windows with low E-coatings and lower solar gain ratings measuring 0.4 or less ensure that the glazing will allow light in but will block the heat. *Solar heat gain coefficients* measure the amount of heat transfer through a material whereas low-emittance coatings applied to glass limit the UV rays from the sun to pass through. For more information on specifying energy-efficient windows for commercial buildings and to work with the interactive Design Façade Tool, visit the website for the Efficient Windows Collaborative and click on "commercial windows" (www.efficientwindows.org).

GO GREEN   Open the Windows

It seems incomprehensible that when the Empire State Building was opened for tenant occupancy back in 1931 that the windows were double hung with sliding sashes. Visions of flying paperwork swirling around the office as someone opened a window on a windy day seem unrealistic. As glass-box buildings emerged during the 1950s with floor to ceiling glass, office buildings were hermetically sealed, relying on ventilation equipment to bring in fresh air. Hermetically sealed interiors contribute to poor indoor air quality when badly designed, and improper maintenance of ventilation equipment can lead to an outbreak of Legionnaire's disease—a deadly respiratory condition during which bacteria enter the lungs. After decades of dealing with poor indoor air quality in office buildings, architects have explored alternatives to a sealed curtain wall.

Designed in 1996 by Ingenhoven Overdiek Kahlen & Partner, the tower built for RWE in Essen, Germany, showcases innovative thinking in bringing natural ventilation into the office interior. According to company principals Christoph Ingenhoven and Achim Nagel, the development of the tower's design came from a desire to "have some linkage with the dream shared by Mies van der Rohe and other architects at the beginning of this century to build truly clear buildings" (Nippon Sheet Glass Co., Ltd.).

The architects designed this building with two "skins" of glass, leaving an airshaft between them. An inoperable exterior glass curtain provides a protective membrane over an interior glass wall that can be opened to bring in fresh air but not the rain. This system, which uses natural ventilation, helps alleviate sick-building syndrome and has launched the consideration of more ecofriendly technologies in today's building structures.

The building for the company headquarters of RWE in Germany, an energy resource company, led the way for green design in 1996 by opening the windows and reducing loads on HVAC systems.

Informative Websites

Americans with Disabilities Act: www.ada.gov

Association for the Advancement of Sustainability in Higher Education: www.AASHE.org

Efficient Windows Collaborative: www.efficientwindows.org

International Code Council: www.icc.org

National Institute of Building Sciences: www.wbdg.org and www.wbdg.org/resources/greenroofs.php

Mechanical Systems for Commercial Buildings

The *mechanical systems* of a building include complex equipment necessary for heating and air-conditioning, *fire annunciation* and suppression, electric power and communications, *conveyor systems,* and plumbing. These systems are designed by mechanical engineers during the earliest stages of planning to ensure equipment is successfully integrated into the building envelope. In addition to determining the equipment needs, engineers plan the network of plumbing pipes, HVAC ductwork, electrical conduits, and data cabling for these systems.

Although the interior designer may not be involved with this part of the process, mechanical systems greatly affect planning of the interior space. For example, closets are needed to house telephone panels and data equipment for computer local area networks (*LANs*), and space is required for mechanical shafts that contain electrical conduits, fire sprinkler pipes, HVAC supply ducts, and plumbing pipes. Moreover, *plenum* spaces—the space between the finished ceiling and the structural slab above—conceal HVAC ductwork, lighting fixtures, fire sprinkler heads, and sound systems (Figures 5.1 and 5.2). Understanding the equipment needs for a building prepares the interior designer for aesthetically integrating these components into the interior planning.

A worker applies a bead of caulking along the seam of an HVAC air duct before installing it in a large building.

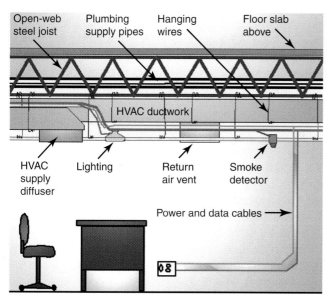

FIGURE 5.1 This drawing illustrates how some of the building's mechanical systems are contained within the plenum of an office building.

FIGURE 5.2 A ceiling grid with removable ceiling tiles allows for easy access to the plenum space where lighting, telecommunications cabling, sprinkler systems, and electrical conduit are hidden from view.

LEARN More

Smart Buildings

The Bahrain World Trade Center towers are the first project in the world to incorporate wind turbines for generating electricity in the design of a commercial building. Influenced by the sails of old fishing dhows, architect Shaun Killa shaped the buildings in a way to capture and direct breezes toward three wind turbines mounted on bridges between the two towers. The turbines are capable of supplying 11% to 15% of the building's power needs, providing electricity for operating mechanical systems like HVAC and lighting. The 1,300 megawatts of power generated by the turbines throughout the course of a year is equivalent to providing enough electricity to light 300 houses for the same amount of time. Additional measures to reduce carbon emissions released through daily operations of the building include *gray water* recycling, site-specific evaporative cooling, a centralized chiller plant for HVAC, and the installation of a glass curtain wall designed to reduce solar gain. Now that is one smart building! For more information, visit the official Bahrain World Trade Center website at www.bahrainwtc.com.

The Bahrain World Trade Center towers are fitted with large wind turbines that supply up to 15% of total energy needs.

Power Distribution Systems

Governed by the *National Electric Code* (NEC) and designed by electrical engineers, the electrical system of a commercial building is quite complex. Not only is there a need for electricity to operate lights and office equipment, but other electrical devices like heating and air-conditioning systems, elevators, fire annunciation systems, and telecommunications equipment must be supported as well. Power from a municipal supply is brought into the building and is connected to a transformer located in the basement or ground floor mechanical room (Figure 5.3). From the main transformer, power is then distributed via a bus duct system that brings electricity to all floors of the building. A panel board on each floor carries the required voltage needed to support the electrical needs for that particular floor. The voltage is distributed through a system of branch circuits (Figure 5.4).

Electrical engineers determine the number of circuits needed per floor to operate all the electrical equipment. A circuit is sized according to how much load is placed on the electrical current, considering, for example, the number of lights or equipment with the same voltage requirements that will be connected to one circuit (Figure 5.5). Voltage is the measured rate of electrical current traveling through wiring at any one time.

Building equipment like elevators and HVAC systems needing more than 120 volts are separated onto dedicated circuits. A dedicated circuit provides electrical current exclusively to a specific piece of equipment and nothing else. Circuit breakers are integrated into the electrical panel boards in case there is an overload caused by having too many appliances or pieces of equipment running off one circuit. The breaker will trip to the "off" position, terminating the flow of electricity and guarding against potential fire or equipment damage. In North America, general office equipment runs on 120 to 125 volts of electricity whereas more powerful equipment like

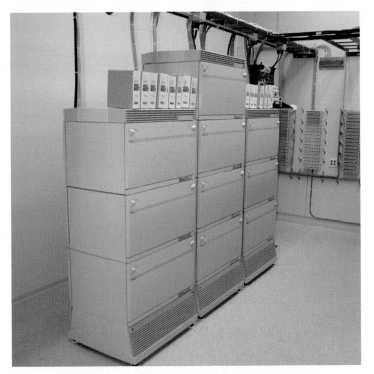

FIGURE 5.3 Electrical equipment, including electrical bus ducts, telephone panel boards, and data cabling devices, is housed in a mechanical room.

GO GREEN Geothermal Energy

A visit to Yellowstone National Park to see the geysers will convince anyone of the potential for *geothermal* energy. Steam from underground springs heated by the earth is used to turn large turbines that generate electricity. Unlike fossil fuels, geothermal energy is clean and sustainable. The Nesjavellir geothermal power station in Iceland operates on four wells. When the underground springs interact with the hot volcanic rock, the reaction produces steam with enough force to produce 2.7 megawatts of power annually. For further information, visit these websites: Energy and Efficiency and Renewable Energy at www.eere.energy.gov and Orkuveita Reykjavík at www.or.is/english.

A geothermal power station in Iceland provides enough steam to operate a power plant for generating electricity.

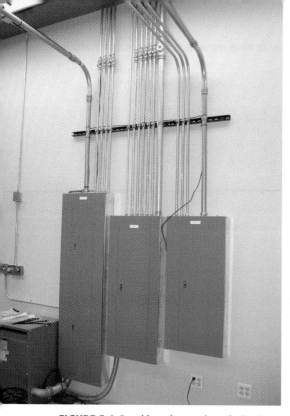

FIGURE 5.4 Panel boards carry branch circuits to different areas throughout a building.

FIGURE 5.5 An electrician standing next to an electrical panel box reads a circuit diagram.

HVAC systems may require 220 volts or 240 volts (Figure 5.6). Telecommunications equipment operates at lower voltages and is run on separate circuits.

Design decisions made by an interior designer include the locations of light fixtures, receptacles, or outlets for office equipment and communication ports. The interior designer works from a furniture, fixtures, and equipment schedule (*FF&E schedule*) and the furniture plan to determine the placement of receptacles that serve furniture groupings, workstations, and equipment (Figure 5.7). Furthermore, the designer must also follow NEC codes, which prescribe minimum distances for locating receptacle outlets. For example, corridors must have an outlet for every 15 feet of

FIGURE 5.6 Appliances list the voltage requirements and the electrical current needs as measured in watts or amperes. Electrical engineers use this information for calculating the electrical loads for a building and for determining the loads placed on branch circuits.

CAUTION

Emergency Generators

To ensure public safety, the National Fire Protection Agency asserts that emergency lighting and fire annunciation and suppression systems remain active in case of power failure. These electrical systems are connected to backup generators that run on batteries or diesel fuel when the electricity is interrupted for any reason. Furthermore, according to the NEC, hospitals are required to have sufficient backup generators (measured in kilowatt hours) to run complex life-saving equipment in case of power outages.

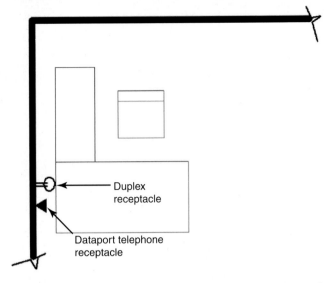

Duplex receptacle

Dataport telephone receptacle

FIGURE 5.7 Furniture plans, which show the locations of equipment and fixtures that require electricity, are prepared by the designer and are used for indicating where communications and electrical receptacles must be located on a power and data plan.

hall length, whereas in other areas (like conference rooms) wall outlets must be no farther than 10 feet apart. These requirements are designed to alleviate the need for extension cords, which are fire hazards and can cause a person to trip over the cord and fall.

Wire management is integral to the successful design of the interior space and, typically, wires for data and communication ports, and electrical outlets are run through plenum spaces (Figure 5.8). Electrical wiring and communications cabling distributed through the plenum space is dropped and run through interior walls or channeled alongside structural columns. These cables connect to junction boxes and data ports, and are concealed behind gypsum wallboard (Figures 5.9 and 5.10).

FIGURE 5.8 Cables in a telecommunications room sit on overhead cable shelves that run throughout the building via the plenum space.

FIGURE 5.9 Electrical wiring is dropped from the plenum space then channeled through openings in the metal studs to reach junction boxes mounted to the side of the metal studs.

FIGURE 5.10 Drywall is cut where access to electrical outlets and communication data ports is needed.

Wiring and cabling in the plenum space of the floor below can be accessed by drilling a hole through the concrete slab and bringing wiring or cabling up to a junction box. This method is frequently used in open office concepts where workstations are not grouped near drywall partitions or columns, and in spaces without a *raised access floor*. For any *core-drilling* project, fire stops designed to contain the spread of smoke or fire must be installed in the plenum area surrounding the drilled core.

In many new construction projects, raised access flooring is installed for wire management. A raised floor is installed over the concrete slab and is designed with removable panels for easy access to wires and cables (Figure 5.11). This type of floor system must be able to support the live loads of people, equipment, and furniture, and the system is covered with quick-release carpet tiles. Furthermore, codes require that all power systems be accessible for repair and maintenance at all times.

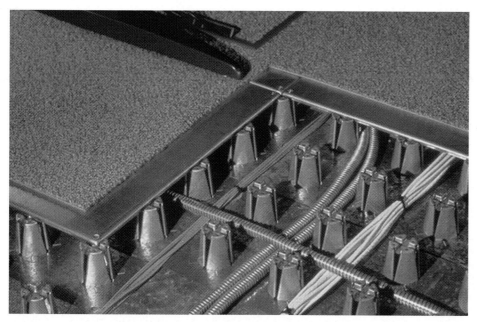

FIGURE 5.11 Raised access flooring makes it easy to get to wiring and telecommunication cables for repair or maintenance.

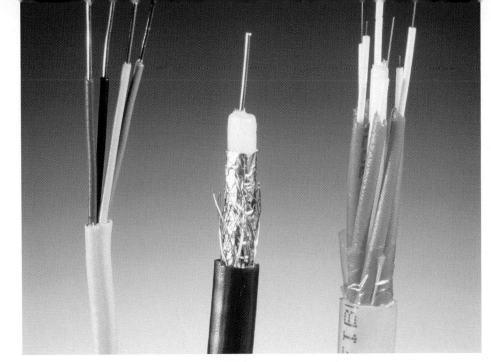

FIGURE 5.12 The development of communications cabling is shown from copper telephone lines, coaxial cabling, and fiber optic cable technologies, from left to right respectively.

Wireless technology has alleviated some of the issues with wire management. Although electricity is still needed to power computers, printers, and scanners, wireless technology eliminates cables otherwise needed for Internet connections and printer networks. The wireless module is powered by electricity, too, and is connected to a centralized LAN for interfacing. In addition, the introduction of fiber optic cabling and its ability to transport information at higher speeds than traditional copper wiring or *coaxial cable* has reduced the amount of trunk lines distributed throughout an office (Figure 5.12).

Heating, Ventilation, and Air-conditioning Systems

HVAC systems are designed to handle the heating and cooling loads of a building. Mechanical engineers must assess the thermal loads of interior spaces to determine the type of system used. The decision to specify prepackaged or stand-alone systems is factored against their performance, energy consumption costs, and environmental impact. All-air systems take air from the outside, which is then heated or cooled, and circulate the air throughout the interior via a network of metal ducts. These systems rely on supply and return ducts to regulate airflow, and thermostats are used to maintain set temperatures (Figure 5.13). The air is heated or cooled via a variety of methods, such as boilers, heat exchangers, or *chillers*.

Prepackaged systems are all-in-one solutions that incorporate heating, cooling, and *ventilation* in one unit that is usually housed on the roof of commercial buildings. Prepackaged systems include *compressors, condensers, evaporators,* supply fans, and filters that force hot or cold air into the central duct system. Chillers and boilers are independent, stand-alone systems used to chill or heat fluids running through a closed-loop system. Chilled or heated water modulates the air temperature as it is directed to an air-handling system for distribution

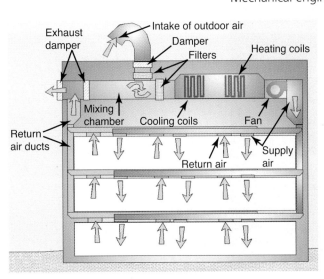

FIGURE 5.13 The basic layout of an all-air system.

LEARN More

Energy Codes and Lighting

The *International Energy Conservation Code*, recently updated in 2009, is the model code that sets requirements for energy efficiency in commercial buildings. The code requires that all buildings larger than 5,000 square feet install automatic shutoff controls for interior lighting. This is a performance code and it can be achieved by installing occupancy sensors programmed to turn the lights out when there is no movement or activity in the room, or by installing timers that shut off all lighting after working hours, weekends, and holidays. The role of the designer when planning for lighting facilitates methods to reduce interior lighting by at least 50% in general areas. This code can be achieved by installing dimmers to reduce the amount of electricity delivered to a fixture, or by a switching plan that activates every other fixture within a room.

Guidelines within the codes set standards for the installation of more efficient lighting by setting limitations on wattage allowances per square foot of space. In addition to allowances per square foot of space, the allowable lighting power density is determined by the type of business. For example,

a fast-food restaurant is allowed 1.4 watts per square foot of space whereas an office is allowed 1.0 watts per square foot of space. To learn more, visit the U.S. Department of Energy website (www.energycodes.gov) and select the link to the section on compliance tools for commercial buildings, or visit the International Code Council website at www.iccsafe.org.

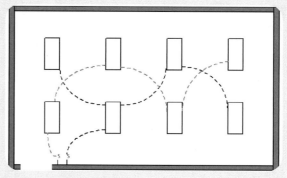

Two switches control half the lights in a room. Keeping half the lights off saves energy.

(Figure 5.14). Air handlers maintain the building ventilation system by bringing in a supply of fresh air that is then distributed throughout the interior. Large, municipally centralized plants produce chilled or hot water that is delivered to the building through a system of underground pipes. Centralized plants are more economical than independent boilers and chillers, because the equipment supplies several buildings in a district rather than one building.

Variable air volume (VAV) systems are small components used to control the amount of airflow and temperature in a distinctive zone (Figure 5.15). Air-handling units distribute air to *VAV boxes*, where thermostats regulate the air (Figure 5.16). These VAV systems are used in buildings where separate temperature controls are needed in the same zone. For example, workers nearest the perimeter curtain wall may experience solar gain in the hot summer months and need more cooling than those working away from the exterior walls.

FIGURE 5.14 Air-handling equipment in this mechanical room will deliver hot or cold air throughout the building.

FIGURE 5.15 Each VAV box installed in the plenum area controls the air temperature in the zone of its associated ductwork.

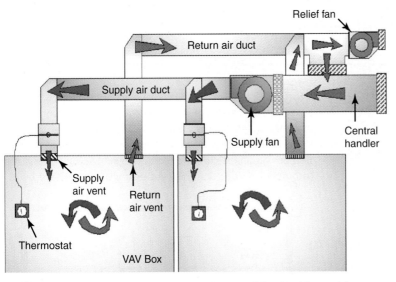

FIGURE 5.16 Air distributed from the air handler is modulated inside a VAV box to control air temperature.

FIGURE 5.17 A man wearing a mask and safety goggles prepares to clean the supply diffuser in an office building. Regular maintenance of ventilation systems promotes good indoor air quality.

Ventilation systems in a building are the most important factor in providing occupants with safe, breathable air. Humidity must be controlled for reducing the growth of disease-causing mold and bacteria. Good *indoor air quality* is achieved with the constant circulation of fresh air and the removal of stale air throughout the building via a duct system. A regular maintenance plan for cleaning HVAC filters and ducts keeps indoor air quality in check (Figure 5.17).

The location of HVAC equipment as well as the network of ducting is determined by the mechanical engineer. Ducts are sized according to the velocity of air forced through the system. Larger ducts are located nearest the air source; ducts get smaller toward the end of the distribution line (Figure 5.18). The interior designer

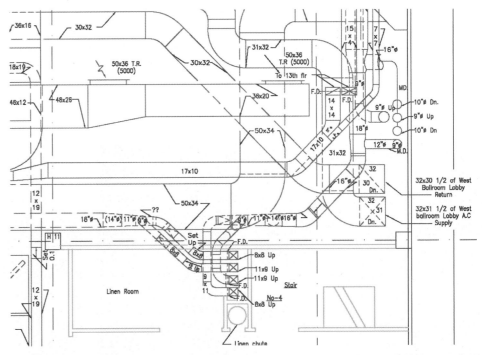

FIGURE 5.18 This HVAC plan shows the distribution of air through a line of ducting sized larger at the source and smaller at the end of the line. Air velocity is greater at the source. Reducing the size of ductwork along the path ensures enough forced air will reach the end.

FIGURE 5.19 Air ducts in this unfinished office space are wrapped with thermal insulation to increase the efficiency of the delivery system.

will review the HVAC plan and will note the locations and sizes of all ductwork in the building to determine possible finished ceiling heights in the space. Ductwork housed in the plenum space must remain clear of obstructions, including housing for light fixtures, sprinkler heads, and communications cabling (Figure 5.19).

Supply registers are distributed throughout the space with consideration of the sources of heat gain or loss. Return grills are usually placed opposite these supply grills to allow the best airflow for maintaining proper ventilation. Thermostats should be located away from sources of heat gain or loss to avoid inaccurate readings of the indoor air temperature. Moreover, thermostat zones must be considered when the interior designer is determining open plan work spaces versus private offices. If a thermostat controlling one zone is placed in a private office, the comfort of those in the open office area will suffer because the thermostat is only getting a reading from the confined area instead of the total space.

Although the engineer plans for the locations of HVAC supply and return grills and thermostats, the interior designer does have a say in how these elements might affect the overall aesthetics of the space. The specification of thermostat control devices and register grills by the interior designer ensures the space has a cohesive and pleasing design. Be aware that the locations of ducting in the plenum space may affect the designer's ability to modulate the shape of the finished ceiling and limit the kind of ceiling system that can be used.

Plumbing Systems

Plumbing systems for commercial buildings operate in the same manner as those for residences. In addition to the basic needs for toilet rooms, kitchens, and laundry rooms, commercial buildings require additional plumbing for utility sinks, drinking fountains, and employee break rooms, and for furnishing water to the building's *fire suppression system* if applicable. As previously mentioned, water is also used by HVAC systems in the building. Plumbing pipes distribute hot and cold water from the central mechanical room to the "wet" areas of the building. Plumbing pipes and drains run through vertical mechanical shafts and are then distributed laterally through the plenum space to reach their specified locations (Figure 5.20). Supply lines pipe hot and cold water, and sloped drains feed into central waste stacks connected to municipal sewer systems.

Plumbing engineers design plumbing systems for large projects whereas architects and interior designers plan the locations for plumbing fixtures. In commercial

FIGURE 5.20 The plenum space in this building shows a network of plumbing pipes that carry *freshwater* to and drainage from the wet areas of a building.

CAUTION

International Plumbing Codes

International Plumbing Codes (IPC) specify the plumbing requirements for commercial spaces based on both the occupancy classification and the occupancy load. In most occupancy types, codes require one drinking fountain per 500 occupants per floor and one utility sink per floor to be used for building maintenance and cleanup. Codes also state the required number of toilets and *lavatories* for male and female toilet rooms based on the occupancy classification and the occupancy load. The following example uses the IPC code to determine the required plumbing needs for "Assembly" occupancy assuming an occupancy load of 552:

- The number of male water closets required is five (552 ÷ 125 = 4.416 rounded up to the next whole fixture).

- The number of female water closets required is nine (552 ÷ 65 = 8.49 rounded up to the next whole fixture).

- The number of lavatories required in each toilet room for male and female is three (552 ÷ 200 = 2.76 rounded up to the next whole fixture).

- No showers or bathtubs are required.

- The number of drinking fountains required is two (552 ÷ by 500 = 1.10 rounded up to the next whole fixture).

- The number of service sinks required is one.

Classification	User Group	Description	Water Closets		Lavatories		Bathtubs and Showers	Drinking Fountains	Other
			Male	Female	Male	Female			
Assembly	A-1	Theaters usually with fixed seats and other buildings used for the performing arts and motion pictures	1 per 125	1 per 65	1 per 200		-----	1 per 500	1 service sink

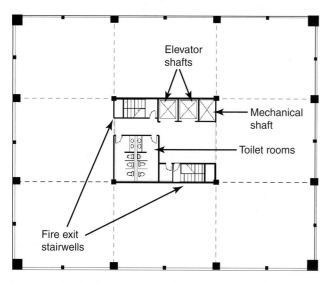

FIGURE 5.21 The fire resistive central core contains elevator shafts, fire exit stairwells, toilet rooms, and mechanical shafts.

Elevator shafts

Mechanical shaft

Toilet rooms

Fire exit stairwells

Gray Water

In 1798, Samuel Taylor Coleridge published the first edition of his poem "The Rime of the Ancient Mariner," and the line "Water, water, every where but nor a drop to drink" was made famous. Although his message refers to sailors on the open sea who are surrounded by undrinkable water, we are experiencing a water shortage on land. Although about 80% of the earth's surface is water, only roughly 1% is drinkable. Because freshwater sources are being depleted, and the technology to desalinate ocean water is energy expensive, new systems developed to recycle water are currently being implemented into the plumbing systems of residential and commercial construction projects. Water from shower drains, sinks, and laundry rooms is harvested then filtered. The water is not safe enough to drink, but harmful detergents are removed, which makes the water suitable for outdoor sprinkler systems and other types of irrigation. Water conservation systems, like using gray water for irrigation, earns points for obtaining LEED certification.

buildings, the stacking of plumbing locations—one above the other on several floors—makes the most sense for providing running water and drainage to toilet rooms, janitorial closets, water fountains, kitchens, and break rooms. These wet areas are usually grouped near the central core of the building nearest the elevator shafts, fire exit stairwells, and mechanical shafts (Figure 5.21), which reduces the distance of running pipes and drains throughout the interior. Furthermore, architects and interior designers are involved in determining the number of plumbing fixtures required per code, the types of plumbing fixtures specified for the job, the layouts of toilet rooms (considering universal access codes), and the interior finish-out of these spaces, including floor and wall materials.

Fire Safety Systems

As promoted by the National Council for Interior Design Qualification, the role of the designer is to protect the health, safety, and welfare of all occupants of a building. Maximizing fire safety is most important in achieving this goal. Fire safety involves the detection, *annunciation*, and suppression of smoke or fire through systems designed by specialized engineers specifically for reducing danger from fire or smoke. The architect or interior designer's role in fire safety is to locate the placement of manual and automatic fire alarms, smoke and heat detectors, and fire extinguishers throughout the interior following performance codes, and to integrate the sprinkler heads within ceiling systems.

Detection is the first step toward saving lives and controlling property damage resulting from smoke or fire. Smoke detectors work in conjunction with heat detectors to warn of possible danger. These two detection systems are used together within the same space to minimize inaccurate readings and false alarms that might otherwise be set off by one detector or the other (Figure 5.22). These devices are integrated into the finished ceiling and are hardwired into

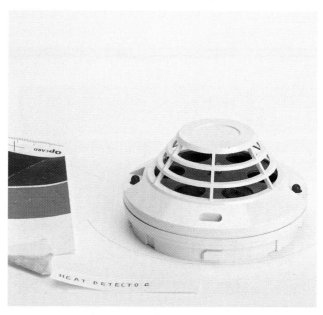

FIGURE 5.22 Smoke and heat detectors are mounted to the ceiling.

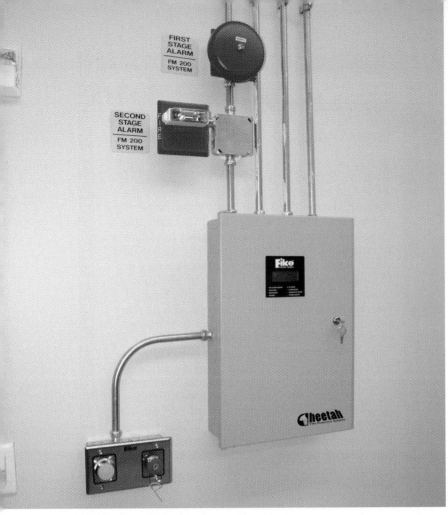

FIGURE 5.23 A fire annunciation system activates fire alarms inside and outside a building, and notifies the fire department of potential danger.

FIGURE 5.24 Manual alarms are required by code for activation by any individual who sees potential danger from fire or smoke.

the main annunciation panel. The annunciation panel, located in the main mechanical room, is the central command for activating fire alarms installed throughout the building (Figure 5.23). Annunciation systems, when activated by either automatic or manual alarms, send a signal to a municipal fire station (Figure 5.24).

Fire suppression equipment is a complex system of tanks, compressors, and pipes designed to supply either water, chemicals, or foam to a network of sprinkler heads installed throughout the building (Figure 5.25). Wet pipe systems use high pressure to dispense water via sprinkler heads into the zone where fire is detected (Figure 5.26). In areas where water could damage property, such as

FIGURE 5.25 Fire suppression equipment is located in the mechanical room of a high-rise building.

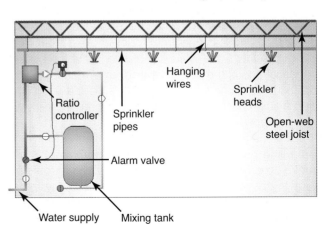

FIGURE 5.26 Fire suppression systems feed water, gas, or foam through a network of pipes housed in the plenum spaces of the interior.

FIGURE 5.27 Alarms like this one, fitted with high-decibel alarms and strobe lights, are distributed throughout the building to meet accessibility codes for audible and visual alarm systems.

books, irreplaceable documents, or computer equipment, systems disperse halogenated hydrocarbon gas or powder into the affected area, thus minimizing damage to these sensitive items. Chemical foams like those emitted from manual fire extinguishers are used to smother the fire and act to contain any toxic fumes or smoke that may be emitted from burning materials like plastics. In addition to suppression systems, special exhaust systems are integrated into the building envelope that are designed to remove smoke from interior spaces, further reducing injury of occupants from smoke inhalation. In most cases, people are killed by smoke inhalation rather than the fire.

Codes require that fire alarms have both visible and audible warning mechanisms. Automatic fire alarms must be located throughout the space where most of the building occupants will be able to hear the alarm or see flashing strobe lights (Figure 5.27). These are placed in corridors, reception and lobby spaces, and inside large conference rooms. Manual fire alarms must be placed near exit doors and mounted on the latch side of the door assembly. Positioning manual alarms nearest exit doors provides a convenient location for activation, because signage throughout the building leads persons to these exits. In addition, fire codes require that fire extinguishers and fire hoses for manual operation be strategically located throughout the interior (Figure 5.28). This fire suppression equipment is often placed in the central core area near fire exit stairwells.

Conveyor Systems

Conveyor systems like elevators, escalators, and moving sidewalks are used to transport high volumes of people efficiently through spaces. All buildings two stories or more are required by universal access codes to include an elevator or lift system. Elevator cabs sized according to weight capacities are raised or lowered throughout

FIGURE 5.28 Required by code, manually operated fire extinguishers must be placed in hallways throughout a building and in those areas where cooking is done.

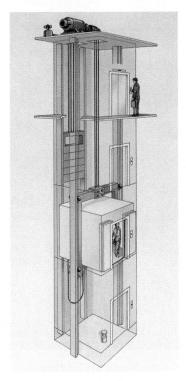

FIGURE 5.29 An electric motor drives cables that set the elevator cab in motion.

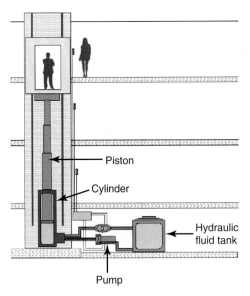

FIGURE 5.30 Hydraulic lifts rely on pistons to raise or lower passenger cabs in elevator systems.

vertical shafts by either cable lifts or hydraulic lift systems. Traction or cable lift systems utilize gears, pulleys, cables, and counterweights to lift and lower passenger cabs between floors (Figure 5.29). Hydraulic systems housed in mechanical spaces located in the substructure or basement of the building use pistons, valves, and hydraulic fluids to push the cab to the upper floors (Figure 5.30).

Elevator shaft enclosures are built to meet specific fire ratings depending upon the building height, occupancy classification, and whether the building has a sprinkler system installed (Figure 5.31). The number of elevator cabs is determined by calculating the peak times of usage for a building based on five minutes total waiting and delivery time. Designers specify interior finish materials for the cabs, select the signage and call buttons, and ensure the interiors meet all accessibility requirements and fire ratings for interior materials. Elevator call buttons must be mounted at wheelchair-accessible heights of 42 inches to the center of the panel, and interior panel buttons are mounted at a minimum height of 35 inches, with the top of the panel not exceeding 48 inches for a front approach or 54 inches for a side approach by persons in wheelchairs. Call buttons must use both visual and tactile identification mechanisms (Figure 5.32).

Escalators and moving sidewalks operate on a closed-loop conveyor system set in motion by large gears. To transport passengers on level steps while going up or down the escalator, steps are designed to flatten as they reach the top and bottom of the loop as they go underneath the conveyor system. Moving sidewalks operate in

FIGURE 5.31 An elevator shaft is finished with fire-rated gypsum wallboard.

FIGURE 5.32 Braille appears next to visual numbers on this elevator control panel.

Long gone are the days when elevator passengers were transported up or down by a uniformed attendant wearing white gloves. Nowadays, passengers are in control with the push of a button, and more advances are being made in the technology of elevator systems. Elevator mechanics is going green. One of the oldest elevator manufacturers, Otis, has introduced a new lift system named Gen2 that reduces operational energy costs up to 50%. The new technology uses flat, polyurethane-coated belts instead of cables and roller guides for raising and lowering elevator cabs safely, without the need for large mechanical equipment. Moreover, the heat generated by the elevator braking system is redirected as energy to supply electricity to run the lighting for the building. This new technology works in buildings up to 30 stories high, and when combined with a ReGen drive system can reduce energy consumption up to 75%.

the same manner as escalators; they rely on gears and motors to maintain a constant rate of motion of the conveyor belt (Figure 5.33). Escalators and moving sidewalks must have closed-in sides, a graspable handrail, nonslip traction, and 24-inch-deep landings, with tactile surfaces at the top and bottom of escalators and moving sidewalks to meet codes. Tactile surfaces give warning to persons approaching or exiting the conveyor systems that movement will begin or end (Figure 5.34).

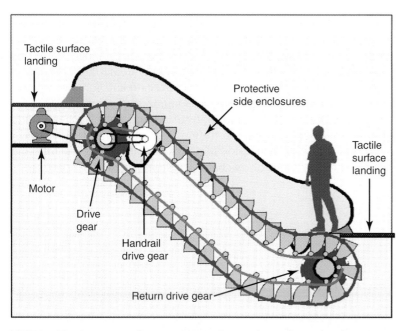

FIGURE 5.33 Motor-powered gears move steps in a continuous loop on escalator systems.

FIGURE 5.34 Codes require escalators and moving sidewalks to have tactile flooring surfaces on all steps, walkways, and landings.

Informative Websites

Bahrain World Trade Center Towers: www.bahrainwtc.com

Energy Efficiency and Renewable Energy, U.S. Department of Energy: www.eere.energy.gov and www.energycodes.gov

International Code Council: www.iccsafe.org

National Council for Interior Design Qualification: www.ncidq.org

National Fire Protection Association: www.nfpa.org

Orkuveita Reykjavikur Geothermal Energy: www.or.is/english

Interior Construction for Commercial Buildings

The interior designer is an important team member who works with engineers, architects, and contractors to see all phases of a commercial project completed—from design schematics to finished construction. In most cases, interior designers are directly involved with the preparation of construction documents needed by the contractor to finish out the interior. These documents detail the locations of interior walls, power and data ports, and the design of ceiling and lighting systems, and provide detailed drawings for all built-ins. In addition, the set of construction documents includes specifications for interior doors, wall and floor finishes, ceiling materials, and furniture layouts. To prepare these documents, the designer must understand the basics of interior construction methods and how walls, ceilings, and floors are affected by the placement of mechanical systems within the space.

Interior finishes in this waiting area feature a polished stone floor, wood ceiling, glass walls, and a custom-designed reception desk.

Wall Assemblies

Interior wall partitions divide and separate spaces into functional zones or areas determined by the interior designer to meet the programming needs of the client. During the space-planning phase of interior projects, the designer decides where to put walls in the building in relation to function, and makes decisions regarding how the partitions should be constructed and how they should be finished. Most partitions are designed to be solid, opaque walls constructed with metal stud framing and enclosed with gypsum wallboard, wood paneling, ceramic tile, or stone veneer. But walls can also be transparent, constructed from glass blocks or *tempered glass* panels. Glazed walls add an interesting aesthetic to the interior and maintain a visual connection between rooms and spaces. Moreover, partitions may be full height or partial height, may run wall to wall, or may be built as freestanding walls within a larger space. Regardless of the type of wall, there are four important considerations the designer must assess when designing a new partition plan: (1) the finish materials in relation to the structural framing, (2) acoustical control needed to reduce sound transmission between spaces, (3) *fire ratings* required by codes, and (4) the overall aesthetic of the wall itself.

Metal studs are used as the supporting framework for gypsum wallboard, stone veneer, wood paneling, or any other cladding material (Figure 6.1). Nonload-bearing walls covered with gypsum wallboard are built from 20- to 25-gauge *galvanized steel* studs. The gauge measures the strength of the steel; lower gauges are actually stronger than higher gauges. For example, a nonload-bearing wall covered with stone veneer will need to be built from lower gauge steel studs ranging from 12 to 18 gauges. Metal studs are manufactured in a variety of gauges and widths. For a gypsum wall, typically 2-½ inch metal studs are used to build an 8-foot-high wall, and 3-⅝ inch metal studs are used to build walls 10 feet or higher. U-shaped metal studs are sometimes called *C-channels* and have holes punched out every few inches to allow electricians to run conduit for junction boxes mounted on the sides of the studs (Figure 6.2, and refer back to Figures 5.9 and 5.10).

The new partition plan in the set of construction drawings informs the contractor of the kind of wall that needs to be built and where. The new partition walls are always

FIGURE 6.1 Metal studs are set in place and are ready to be enclosed with gypsum wallboard.

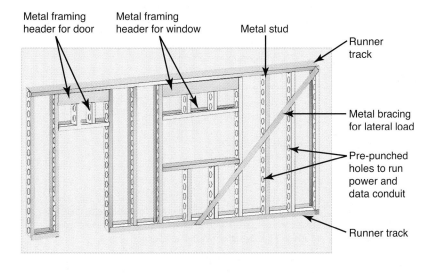

Metal framing header for door

Metal framing header for window

Metal stud

Runner track

Metal bracing for lateral load

Pre-punched holes to run power and data conduit

Runner track

FIGURE 6.2 This drawing shows the structural framing for a wall with openings for a door and windows.

Fire-Rated Partitions

Maintaining fire safety in commercial interiors is the driving force behind the construction of interior wall partitions. Metal studs and gypsum wallboard are inherently fire resistive and noncombustible, making these materials good choices for wall assemblies. Partition wall assemblies must be constructed to meet specific fire ratings determined by the occupancy classification for the project and the location of the partition wall in relation to *exit access* and exits. Specifically, in locations where there is the potential for an increase in fire hazard, performance codes require fire-rated wall assemblies. For example, corridors leading occupants to exits or exit access must maintain a one-hour fire rating, and two-hour-rated partitions are required for exit stairwells and other vertical shafts like those used for conveyor systems. Fire-rated wall assemblies are specially constructed to withstand structural failure from exposure to smoke or fire for a specific amount of time. Based on laboratory testing, a one-hour fire-rated wall assembly will resist these threats for up to one hour's time, a two-hour fire rating resists damage or infiltration for up to two hours, and so on. The architect and interior designer are responsible for specifying partition types on a new construction plan that meet current codes. Partitions are labeled on the plan and are keyed to a legend with specific details for their construction. Like most interior walls, metal studs and gypsum wallboard are the primary materials used to construct fire-rated wall assemblies; however, CMUs are often used in the construction of fire exit stairwells. The following drawing indicates one example of how a one-hour fire-rated wall assembly can be built. The chart indicates other methods for achieving fire-rated assemblies.

Fire Rated Partitioning for Commercial Construction	
Fire Rating	**Construction Description**
1 Hour	Type X gypsum wallboard ⅝-inch thick, mounted on each side of 2½-inch metal studs. Partition is built from the floor slab to the underside of the slab above.
2 Hour	Type X gypsum wallboard ⅝-inch thick, mounted on one side of 2½-inch metal studs with ⅝-inch thick gypsum wallboard and 1-inch thick Type X gypsum wallboard applied on opposite side. Partition is built from the floor slab to the underside of the slab above.
2 Hour	Two layers Type X gypsum wallboard each layer ⅝-inch thick, mounted on each side of 2½-inch metal studs. Partition is built from the floor slab to the underside of the slab above.
3 Hour	Type X gypsum wallboard 1-inch thick, mounted on each side of 2½-inch metal studs with ½ inch thick Type X gypsum wallboard applied on each side. Partition is built from the floor slab to the underside of the slab above.

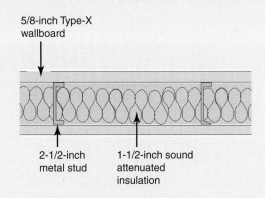

Plan Section View

Type X wallboard ⅝ inches in thickness is applied to each side of 2½-inch metal studs that are located every 24 inches on center.

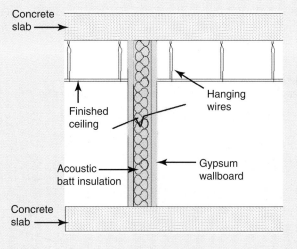

Section View

A partition built from the top of a concrete floor slab to the underside of the slab above creates a fire block in the plenum. Sound attenuation insulation 1½ inches thick fills the wall cavity.

dimensioned from a point of existing or fixed structure, like columns or core walls, and each partition is keyed corresponding to a legend, with complete descriptions for its construction. For interior walls, metal studs are placed every 16 inches or 24 inches on center, depending on the weight of the final finishing material. Curved partition walls are created by spacing the studs closer together on spliced *runner tracks* following the radius of the curve (Figure 6.3). Two layers of ¼-inch-thick flexible gypsum wallboard is applied over the studs for dimensional stability (Figure 6.4). For tight curves, wetting the gypsum wallboard before attaching it to the stud framing gives it more flexibility.

After the stud walls are in place, electrical conduit is run to junction boxes placed on wall locations according to the power and data plan. When needed, batt insulation (providing thermal and acoustical control) is added before the wall cavity is enclosed (Figure 6.5). Gypsum wallboard is secured to the metal studs with screws, and the joints between panels are taped and bedded to give the interior wall an even and seamless surface. As mentioned in Chapter 3, the process of taping and bedding requires great skill. First, a thin layer of joint compound is applied over the seams using a 4-inch putty knife. Specially made paper tape is then applied and allowed to dry thoroughly before a second coat of joint compound is applied over the tape, extending beyond its edges (Figure 6.6). After this dries, a finish coat of joint compound is applied, with the skill of the worker feathering the edges to hide any visible remains of the seam. After the tape-and-bed job has thoroughly dried, the joints are then "sanded" with a wet sponge to give the wall its smooth finish.

LATH AND PLASTER PARTITIONS

Curved walls or architecturally designed elements exceeding the capabilities of flexible gypsum wallboard require a more malleable finishing process using lath and plaster. The flexibility

FIGURE 6.3 A worker forms the runner track for a curved wall.

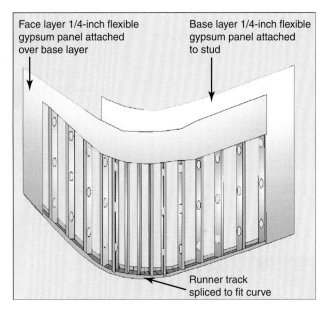

Face layer 1/4-inch flexible gypsum panel attached over base layer

Base layer 1/4-inch flexible gypsum panel attached to stud

Runner track spliced to fit curve

FIGURE 6.4 Gypsum wallboard is curved over steel studs.

FIGURE 6.5 Workers set a panel of gypsum wallboard onto a metal stud wall.

FIGURE 6.6 Tape and bedding finishes the wall for a smooth surface.

FIGURE 6.7 The staircase in this southwestern-style hotel features curved handrails formed from metal lath and plaster.

of the lath and plaster allows for tight curves and decorative surface treatments (Figure 6.7). A meshlike metal or plastic lath is the substrate for layers of plaster. *Gypsum plaster* is applied over the lath in three stages. The first layer is a thickly applied ½-inch scratch coat, the second layer is a ¼-inch-thick leveling coat, and, last, a smooth finish coat about ⅛-inch-thick fully levels out the wall to conceal any evidence of the lath.

FIGURE 6.8 A curved glass block wall provides both light and privacy for this conference room wall.

GLAZED PARTITIONS

Glazed partitions are walls made entirely or in part of glass. The most familiar type of glazed walls are those made from glass block (Figure 6.8). Glass blocks are laid in courses; the first course is laid on a sill plate and then each course is stacked one on top of the other, with reinforcing rods in between (refer back to Figure 3.8). As a building material, glass block is strong enough to support its own weight, but cannot be used as a load-bearing wall. Glass blocks are easily fitted into gypsum wall assemblies as accents or are left freestanding using specially designed glass block end caps. The blocks are vacuum sealed and carry a fire rating from 45 to 60 minutes, and are also good for controlling sound (maintaining a minimum 40 *sound transmission classification* [STC] rating).

Glass partitions are quite common in the design of office spaces, retail storefronts, and businesses (Figure 6.9). Glass can be molded into curved panels, etched with images or designs to make the glass opaque, or filled with a sheet of liquid crystals that change from transparent to opaque when activated by an electric current (*LCD glass*). Codes require glass partitions to be made from tempered or safety glass, which is stronger than regular glass. Moreover, unlike regular *float glass*, safety glass breaks into small pebblelike fragments instead of sharp-edge shards. Tempered glass is manufactured to

FIGURE 6.9 Glass wall panels are held in place with aluminum framing channels concealed in the floor and ceiling.

the exact height and length specifications where the panel will be installed. After tempered glass is made, it cannot be cut. The thickness of tempered glass panels is determined by the height and span of the wall. Glass wall panels are installed in an aluminum support, framed into gypsum wallboard, or placed by securing the glass into runner tracks embedded in the ceiling and floor (Figure 6.10).

Interior Doors

Interior doors for commercial spaces are selected for many functions, including the ability to maintain visual privacy when needed, control acoustics within a space, secure access to other areas, or prevent the spread of smoke or fire from

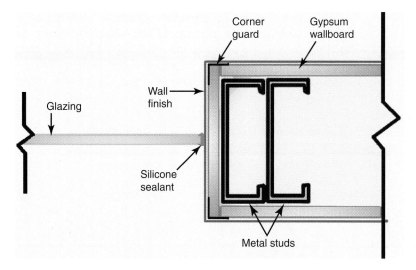

FIGURE 6.10 This detail shows how a frameless glazed panel abuts the end of a gypsum partition. A small amount of silicone sealant is applied at the joint to stabilize any movement from vibrations.

FIGURE 6.11 Combination wood and glass doors complement the glazed corridor walls in this office interior.

one area to another. Before selecting doors, performance and aesthetics are considered: How will the door hold up after repeated use? Should the door be glass, metal, wood, or a combination of materials? Which door meets universal access codes? These decisions are made when developing the aesthetic concept of the space and during the code review process (Figure 6.11).

Universal access codes require that door widths have a clear opening of 32 inches measured from the jamb side to the surface of the opened door. Heights must be a minimum of 80 inches clear of any obstruction like automatic door closers (Figure 6.12). Accessibility codes also require that doors have easy-to-operate, lever-type hardware mounted from 34 to 48 inches above the finish floor (AFF) (Figure 6.13). The new partition plan in the set of construction drawings shows locations for all doors and is keyed according to a separate schedule (refer back to Figure 1.37). The occupancy load for each space determines whether the door must swing in to the room or out into the corridor per code. In most cases, in rooms with an occupancy load of 50 or more, doors must swing out of the room in the direction of exit travel. In addition, doors opening into the corridor must not protrude more than 7 inches into the corridor pathway (Figure 6.14).

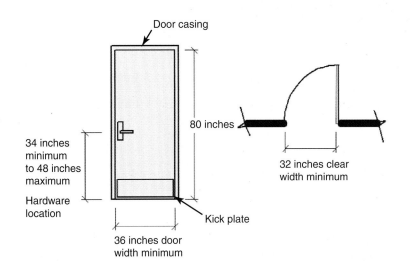

FIGURE 6.12 This drawing indicates door measurements and other requirements for meeting universal access codes.

Ceilings

Interior space is modulated by its enclosing walls, ceilings, and floors, and the configuration of these elements contributes not only to the aesthetic concept, but also to human behavioral responses as well. Ceilings play a significant role in how people feel in a particular space. If the ceilings are too high, the occupant may feel small and insignificant; too low, people may feel threatened or confined. Certainly, from time to time, interior designers intentionally manipulate the scale of a ceiling to evoke specific responses, create dramatic impact, or challenge the notion that a ceiling is simply a covering over our heads (Figure 6.15).

When preparing a *reflected ceiling plan* for inclusion in the set of construction drawings, the interior designer has more to consider than ceiling height alone. Materials contribute significantly to the complete aesthetic makeup of the space and affect the acoustics within. Ceiling construction and finishing materials are subjected to strict codes to ensure structural integrity and fire safety. Moreover, consideration for how ceiling components interact with building systems housed in the plenum, like HVAC

FIGURE 6.13 This lever-type door handle meets the requirements of the Americans with Disabilities Act for its ease in operation.

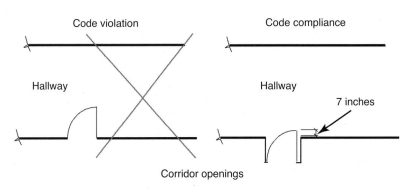

FIGURE 6.14 Doors opening into corridors must not protrude more than 7 inches according to code.

Fire-Rated Doors

Interior designers are responsible for specifying doors that meet all current codes, including fire safety. In egress areas like fire exit stairwells, fire-rated doors must carry the same fire rating as the wall assembly where it is installed. To earn a fire rating, the total door assembly, including the door panel, its frame, and hardware, must perform as a unit to resist the spread of fire or smoke for a designated period of time (usually from 45–90 minutes). Fire-rated doors are constructed from noncombustible materials, and an insulating core minimizes the transfer of heat through the door panel. Hardware must be easy to operate with slight force, and all fire-rated doors must be self-closing or installed with automatic closers. A fire door left open adds to the risk of injury and is ineffective in compartmentalizing smoke and fire.

Self-closing fire doors with fire-rated glass inside this hospital aid in compartmentalizing areas to stop the spread of fire or smoke.

ducting, sprinkler pipes, lighting, and communications cabling, is important. The type of ceiling specified for commercial interiors must also comply with codes that require adequate access to reach mechanical systems for repairs.

Ceilings for commercial interiors made from gypsum wallboard offer a smooth and seamless surface and are constructed with the same structural framing used for building interior partition walls. Metal stud framing is installed according to the shape of the final ceiling design and provides the structural support for the finishing material. Metal studs are secured to the overhead concrete slab, steel beams, or trusses.

FIGURE 6.15 Undulating ceiling panels appear to float above the furnishings inside this conference room.

CAUTION

Fire Safety and Interior Finish Materials

Fire ratings for flooring materials, ceilings, and walls are based on flame-spread testing. Depending on the occupancy classification of the project, national fire safety codes will determine whether a class A, B, or C rating is required. Materials are tested and awarded a fire rating (these same ratings apply to wall finish materials). Fire ratings are determined by how fast flames will spread across the surface, and range from 0 to 100; the lower the number, the more resistive the material. Class A ratings are given to materials having a flame spread of 0 to 25. Class B are those with flame spreads of 26 to 75, and class C ranges from 76 to more than 200. The architect or designer must consult the product specifications to ensure the specified product meets current codes.

Ceilings are an important element in keeping smoke and fire from spreading through the plenum space and throughout the rest of the building. Materials used in ceiling construction must meet flammability standards based on laboratory testing.

Visual Selection				UL Classified								
Edge Profile	Suspension Detail Dwg. Pgs. 226–228	Item No.	Dimensions	Acoustics NRC CAC	Fire Rating	Light Reflect	Sag Resist	Anti-Microbial	Durable		Recycle Program
ENDURA Tegular											
9/16" Beveled Tegular	31–34, 54	639 639M	2' x 2' x 3/4" 600 x 600 x 19mm	0.65 35	Class A	0.84	HumiGuard+	BioBlock+	Impact Scratch		Yes
15/16" Angled Tegular	8	638 638M	2' x 2' x 3/4" 600 x 600 x 19mm	0.65 35	Class A	0.84					
	8	640 640M	2' x 4' x 3/4" 600 x 1200 x 19mm	0.65 35	Class A	0.84	Standard				

Performance Selection Dots represent highest level of performance.

Suspension Systems

15/16" Prelude®   9/16" Interlude®   Silhouette® Bolt-Slot   Sonata®   Suprafine   Trimlok® Screw-Slot

Item 639-CAC 33 on 9/16" Interlude, Sonata, Suprafine

The fire ratings for lay-in ceiling tiles and the suspended ceiling grid are listed on the product specifications sheet.

Full code compliance determined by the occupancy classification and ceiling location in the building is required when specifying ceiling materials. For example, ceilings located nearest exits or exit accesses must have a class A fire rating. Fire ratings for ceiling products are listed on the specifications sheet.

FIGURE 6.16 This conference room features two distinctive ceiling treatments. The suspended ceiling allows for easy access to integrated systems contained in the plenum whereas the dropped gypsum wallboard panel reflects the sound of voices during meetings.

Wallboard is attached to the metal studs with screws placed every 12 inches to ensure that the weight of the drywall does not deflect over time from the constant pull of gravity. The wallboard is then taped and bedded, primed, and painted. The hard surface of gypsum wallboard reflects sound, making it a suitable material inside large conference rooms where sound amplification is needed. Sound is reflected off gypsum wallboard ceilings, so conversations around the conference table are easily heard by all participants (Figure 6.16).

In open office spaces, sound amplification is less desirable than inside conference rooms, and acoustical tile ceilings offer the best solution for noise control. Acoustical tiles installed in a suspended ceiling system also provide easy and quick access to the plenum space to service light fixtures, smoke alarms, sprinkler systems, plumbing pipes, and HVAC ductwork. These mineral fiber panels are most commonly used in commercial interior projects because of their fire ratings, acoustical properties, and low cost (Figure 6.17). The panels have tiny perforations that provide acoustical control. Suspended ceiling systems are practical and easy to install by attaching a lightweight metal grid to the structural components of the ceiling joists or concrete slab with hanging wires. In addition, a suspended ceiling grid system is used to support more than just mineral fiber ceiling tiles. Many vendors supply metal, glass, plastic, and wood panels fitting suspended grid systems in sizes of 2 by 2 feet or 2 by 4 feet (Figure 6.18).

The concept of leaving a ceiling open and exposing the mechanical equipment and structural components above it originated with the movement toward a "factory" aesthetic during the 1980s (Figure 6.19). Open ceilings have long been present in the design of modern interiors by being either nonexistent or merely alluded to by the presence of a suspended metal grid (Figure 6.20). Disadvantages of using open ceilings include a loss of the element of human scale, the lack of acoustical control, and the provision of a more industrial feel to the space.

FIGURE 6.17 A suspended ceiling with acoustical tiles are featured in this bank lobby.

Floors

In all commercial construction projects, the structural components of floor systems—including open-web steel trusses, metal decking, and poured concrete, or concrete beams and prefab concrete panels—provide the necessary support for any floor finishing material. Guidelines for the installation of finishing materials like stone, ceramic tile, carpet, or resilient flooring are detailed in the specifications issued by the manufacturers of these flooring products. It is important for the architect or interior designer to include these specifications in the package of construction documents distributed to the general contractor and the installer. Improperly installed flooring materials may void the manufacturer's warranty for product performance, and an unevenly installed floor could violate accessibility codes or lead to personal injury.

FIGURE 6.18 A lightweight suspended ceiling grid supports a wooden lay-in ceiling.

FIGURE 6.19 An open ceiling in this office space leaves HVAC ductwork and electrical conduit exposed as a design feature.

FIGURE 6.20 An open-grid suspended ceiling provides an element of human scale by lowering the connection between slabs.

Universal access codes require that flooring materials in commercial spaces comply with prescriptive codes relating to the transitioning between one flooring material to another, and set the maximum allowable thickness for all flooring materials. Floor finishing materials like carpeting, hardwood strips or planks, wood veneer or plastic laminate strip flooring, ceramic or stone tiles, vinyl, cork, synthetic rubber, and linoleum must also meet flammability requirements prescribed by code.

CARPET

There are many advantages of specifying carpeting for commercial interiors. Carpet absorbs sound, helping maintain noise reduction within spaces. It also adds pattern and texture to the interior, and it is more comfortable to stand on for long periods of time than other flooring materials. The leading type of carpeting specified for commercial interiors is in the form of carpet tiles. Unlike broadloom carpeting, which is most common in residential applications, carpet tiles are manufactured with an attached backing material to eliminate the need for a carpet pad. Carpet tiles are easily replaced in case of excessive wear and are best for use in buildings with raised access floors. Carpet tiles are installed with a quick-release glue, or even Velcro, so they can be easily removed for accessing cabling and electrical conduits in raised access floor systems (Figure 6.21).

Accessibility codes require that all carpets installed in commercial spaces have a low, level surface not exceeding ½ inch in thickness, making it easier for persons in wheelchairs or with other mobility problems to move about the interior without difficulty. Keeping a safe environment for all, codes require carpeting to have a firm backing and be securely attached to a substrate material with no loose edges, which can cause people to trip and fall. Codes allow for area rugs to be used in commercial spaces as long as their edges are securely fastened to the floor and covered with a beveled transition strip.

WOOD LAMINATE FLOORING

Wood flooring has a warm aesthetic with its rich wood graining and color, and it is suitable for commercial interiors as long as the material meets commercial class fire ratings. Wood flooring specified for commercial applications is an engineered product designed to withstand heavy traffic while meeting the minimum requirements for fire safety. Wood flooring can be installed in two ways: applied directly over the concrete subfloor with adhesives or by using a floating floor system. Floating floor installations are those in which floorboards connected to each other with tongue-and-groove joints are laid over a polyethylene underlayment.

FIGURE 6.21 Carpet tiles in gray and tan create a dynamic visual feature in this office space.

STONE, CERAMIC, AND QUARRY TILE

Wherever water is used in commercial interiors—kitchens, break rooms, toilet rooms, utility areas, and so forth—codes require that the walls and floors have a continuous impervious surface that holds up to constant cleaning. Ceramic or quarry tiles are the best solution for these wet areas because they are inherently water resistant, nonabsorbent, easy to clean, and hygienic.

All tiles specified for commercial flooring must have a slip-resistant finish to comply with codes. Stone tiles are usually installed in high-traffic areas like public lobby spaces and, unless treated with a nonslip surface, cannot be installed as flooring in wet areas (Figure 6.22). Stone or tile flooring is installed with a mortar bed over the concrete slab and the gaps are filled in with grout. Subflooring other than a concrete slab needs extra reinforcements to support the weight of the tile or stone.

TERRAZZO

Terrazzo is the most commonly used flooring material for heavy-traffic areas like airports and shopping malls (Figure 6.23). Made from marble or granite chips and cement, this low-maintenance floor material is installed by pouring the self-leveling mixture over concrete subfloors and polishing it to a smooth surface after the cement has dried. Thin metal bands are used to lay out patterns or shapes before pouring. Terrazzo is inherently a green product because it is made from recycled aggregates, including glass, and does not contain any volatile organic compounds. For more information on this green product and available LEED credits for specifying this material, visit the National Terrazzo and Mosaic Association website (www.ntma.com) and link to their sustainability and LEEDS information under the "Terrazzo" information menu on their homepage.

RESILIENT FLOORING

Resilient flooring materials are pliable, providing a more cushioned floor than hard-surface flooring like wood, stone, or tile. Manufactured from a range of man-made materials like vinyl or synthetic rubber, and natural materials like linoleum and rubber, resilient flooring materials are suitable for heavy-traffic areas like

LEARN More

Specifications and Manufacturer Warranties

Specifications are an important component of the construction documents package and are provided by manufacturers to protect product warranties. These specifications contain details on the installer's responsibility, the storage and handling of the materials before installation, and how to prepare the installation site to ensure maximum performance of the product. The architect or interior designer is responsible for providing the general contractor and/or installer with complete manufacturer specifications to do the job correctly. Visit the Armstrong website (www.armstrong.com/pdbupimages/179764.pdf) to view a sample of a specification for installing a wood floor.

FIGURE 6.22 Polished marble flooring is used in the lobby of this high-end hotel.

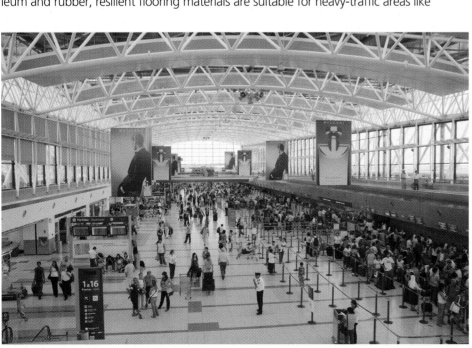

FIGURE 6.23 A terrazzo floor holds up against heavy traffic, is easily maintained, and offers flexibility in design and color.

113

FIGURE 6.24 A vinyl floor provides a durable and practical finish inside this employee break room.

corridors, classrooms, and hospitals (Figure 6.24), because they are economically priced and easily maintained. Resilient flooring materials specified for commercial projects must meet required fire ratings according to the occupancy classification of the project.

In hospitals and health care facilities, antimicrobial flooring must be specified to ensure that germ-free conditions are maintained. Self-leveling poured-in-place floors eliminate bacteria-collecting seams. The material is poured 4 inches or more from the floor onto the side walls to eliminate crevices and to ensure a hygienic environment. Microbial sheet vinyl is also specified for these occupancies as long as the seams are heat welded to maintain an aseptic monolithic floor covering.

Static control vinyl flooring is available for those areas with equipment sensitive to static electricity, like computer centers, biotechnology laboratories, clean rooms, and electronics manufacturing plants. It is important to specify a conductive vinyl adhesive with these flooring materials to maintain the resistance of the product. A greener option, like conductive rubber flooring made from recycled rubber, is also available. Static control materials enable static electricity to pass through the flooring and adhesives into the ground, thus eliminating damage caused by static electricity.

LEED the Way

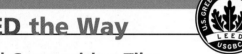

Vinyl Composition Tiles

Vinyl composition tiles have come a long way since they were first introduced in the 1950s and included asbestos in their manufacture. Producers of vinyl tiles have introduced measures to gain LEED points by introducing recycled materials in their products and by using low-VOC adhesives. Today's vinyl composition tiles are made of recycled content and include limestone, PVC resin, plasticizers, and pigments. These tiles are good options for floor coverings in light industrial, institutional, and educational facilities, because they are durable, easy to maintain, impact and stain resistant, and low cost. Check the manufacturer's specifications to determine the LEED credits that apply. Points are earned for those products made from 10% to 20% recycled content (LEED MR 4.0: Use materials with recycled content such that the sum of the postconsumer recycled contents plus one-half of the preconsumer [postindustrial] content constitutes at least 10% or 20% of the total value of the materials in the project) and are installed with low-emitting adhesives and sealants to maintain good indoor air quality (LEED EQ 4.1: All adhesives, sealants, and sealant primers must comply with the requirements of the South Coast Air Quality Management District Rule #1168. Aerosol adhesives must meet the Green Seal Standard for Commercial Adhesives GS-36). For more information, visit the following link to read a summary of LEED credits: www.usgbc.org/ShowFile.aspx?DocumentID=684.

Millwork

Millwork construction for commercial office spaces, like custom-designed built-ins, bookcases, reception desks, and toilet room vanities, is part of the finish carpentry that completes a design aesthetic not offered through stock cabinetry. In commercial spaces, millwork also includes internal staircases between floors of the same tenant, ceiling moldings, wall paneling, baseboards, and door and window trim. Reception desks in offices, hotels, or bank lobbies, and the cash wrap in a retail store are examples of custom-designed millwork for fabrication (Figure 6.25). These millwork elements lead the traditional, contemporary, or transitional design concept planned early during the schematic design phase. Concept drawings are sketched and final working drawings are prepared as part of the construction documents package (Figure 6.26).

Milled trims, moldings, baseboards, and wall paneling designed for commercial interiors must comply with codes that address interior finish materials. These architectural details act as fuel in case of fire and, depending on the material, can emit toxic chemicals when ignited. All fire-treated, but still combustible, trims are required by code to be limited in application to 10% of the combined area of walls and ceilings for the space where installed. The rule excludes wooden handrails where required by code, such as in hospital corridors, nursing homes, and senior daycare centers. Noncombustible trims like those molded from gypsum plaster or prefabricated composite moldings must have a class C or better fire rating and are unlimited in percentage of application.

Stairs

Fire stairs are designated means of egress, and the code requirements for their construction are most restrictive. Fire escape stairwells are designed along with other central core

FIGURE 6.25 A custom-designed reception desk features combinations of wood and stone.

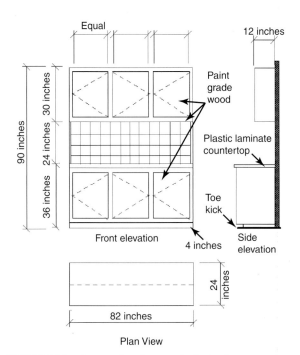

FIGURE 6.26 Concept drawings set the basics for a custom designed storage unit.

items like toilet and mechanical rooms, and elevator shafts. Wall assemblies enclosing fire stairs must have a minimum one-hour fire rating or higher, determined by the occupancy classification. Minimum stair widths are 44 inches, although some occupancy classifications require 60 inches depending on the occupancy load. Stairs must have handrails on both sides of the enclosure (Figure 6.27). Doors must open into the fire stairwell without obstructing the path of those descending the staircase and must be equipped with panic hardware. Exhaust ventilation systems are installed to remove smoke from the fire stairwells and make them suitable for areas of refuge.

Internal staircases that link the operations of the same business from one floor to adjacent floors do not need to comply with the codes for fire stairs. However, all staircases must comply with codes that set the size for tread depth, riser height, stairwell width, and handrail height.

FIGURE 6.27 The design and interior finish-out of fire exit stairwells are governed by strict codes to ensure a safe evacuation for the building's occupants.

LEARN More

Areas of Refuge

In emergency situations, the activation of the fire alarm signals all elevators to return to the ground floor as a safety precaution. For persons with restricted mobility, like persons in wheelchairs or those who use walkers, getting down the fire stairs is problematic and they pose dangerous obstacles to the safety of others. *Areas of refuge* are planned spaces where people can go and wait for rescue. They can be designed as special rooms or designated areas within the fire stairs enclosure. In the 160 story high-rise, Burj Khalifa, evacuation using fire stairwells would exceed an hour, which is not a practical solution. The architects, Skidmore Owings and Merrill, provided each floor with a specially built area of refuge complete with two-hour fire-rated walls and special ventilation. Persons with impaired mobility or disabling health problems are sheltered in these areas of refuge until help arrives. These people are then evacuated via special water-resistant elevators operating on emergency generators.

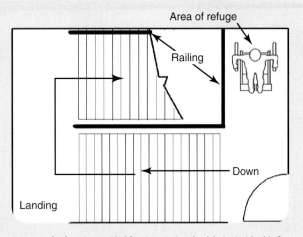

An area of refuge is provided for persons in wheelchairs inside this fire stairwell.

Acoustics

Acoustical objectives for commercial interiors vary from project to project. Concert halls, subway stations, or busy restaurants all require different types of sound control to contribute toward the human experience of being in these spaces. A good example of acoustical manipulation to create a mood occurs in restaurant design. In establishments where meals are cheaply priced, noise levels are kept high by limiting sound-absorbing interior finish materials. Elevated noise levels detract people from lingering in long conversations and ensure tables will be turned over to new customers more frequently. On the other hand, a nice restaurant with higher priced food might have a relaxed and quiet atmosphere to foster a more leisurely meal; the tables might be turned over less often than the inexpensive restaurant, but the profits are made up in menu pricing.

The role of an interior designer is to determine sound sources, activities taking place within a space, background noise, the dimensional aspects of a space related to room size, and room finishes for setting acoustical objectives for a commercial project. Noise is generated by a variety of sources, including the sound of forced air through HVAC ductwork; equipment such as personal computers, telephones, or copy machines; and conversations, group activities, and meetings held in open-office spaces. Outside sources like traffic noise from cars and airplanes, playgrounds, or parks infiltrate the interior. After a thorough analysis of acoustical needs, a designer specifies materials and finishes that either absorb or reflect sounds, and control sound in and in between spaces.

Space planning helps control sound transmission between spaces by locating quiet areas away from noisy ones, or, more specifically, by separating passive areas from active ones. Controlling acoustics between adjacent spaces is achieved through interior wall partitioning, ceiling materials, and construction by specifying finish materials that absorb sound (Figure 6.28). The ability of sound to travel through barriers like walls, doors, or ceilings is rated according to STC. STC ratings are numeric; the higher the number, the more sound is blocked from passing through the wall, floor, or ceiling assembly. An STC rating of 25 allows normal speech to be heard through the sound barrier, whereas assemblies with an STC rating of 46 to 50 eliminates loud noises.

To keep sound from traveling through the plenum, barriers are installed to control sound leaks and to act as sound blocks. Sound travels up through the ceiling into the plenum and then down through the ceiling of the adjoining space. Sound leaks also travel through the HVAC ductwork and through partition walls wherever there are openings, like those for electrical outlets and data ports (Figure 6.29).

Constructing partition walls with high STC ratings is the best precaution against sound transmission occurring between spaces. Building a partition wall from the concrete slab to the underside of the top slab blocks sound from breaking through the plenum to adjacent spaces. A drywall partition constructed in this manner achieves the highest STC rating. Adding acoustical batt insulation inside the wall cavity during construction achieves additional sound-blocking control. Where maximum sound control is needed, the partition wall is built with two layers of acoustical gypsum wallboard (Figure 6.30). Sources of sound leaks must be sealed with acoustical sealant, and doors must be specified with a comparable STC rating that maintains the STC rating of the partition (Figure 6.31).

It is important to note that sound-blocking materials are not necessarily sound absorbers. Materials should be specified for both sound-absorbing and

FIGURE 6.28 Batt insulation is encased inside the wall cavity to provide acoustical control between partitions.

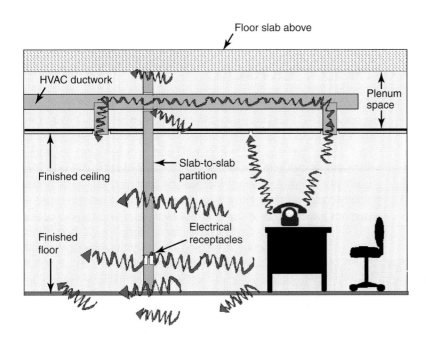

FIGURE 6.29 This diagram reveals the sources of sound leaks from one space into an adjacent space.

sound-reflecting qualities. For example, ceilings made from drywall amplify sound and are commonly placed above conference tables to allow sounds to "bounce" off the ceiling and project voices so everyone can hear the conversation at the table.

Essentially, sound-reflecting surfaces are necessary to enhance intelligibility of speech, to hear music projected from the stage into the audience, or to experience the deafening excitement of spectators at a basketball game. The amount of sound absorption depends upon the thickness, density, and porosity of the material used. Interior finishes made with sound-absorbing properties like acoustical fiber tiles, perforated metal panels, or an engineered wood ceiling installed with air space between the slats are designed to reduce noise. Carpeting, upholstered furniture, walls covered with acoustical panels or sound-rated vinyl wall covering, and soft window treatments control noise, eliminate reverberation, and preserve speech privacy.

Products are rated with a *noise reduction coefficient* (NRC) that is expressed numerically based on the material's ability to absorb sound; the higher the number, the better the sound absorption. An acoustical tile ceiling is the best example of a material specified according to its NRC rating. Acoustical tiles in private offices should have an NRC rating of 0.60 or higher. Ideally, ceilings in open plan offices should have an NRC rating of 0.80 or higher to absorb sound (refer to the specification in "Caution: Fire Safety and Interior Finish Materials" "on page 110"). Additional ratings are given to ceiling finish materials that act as sound blockers as indicated by its *ceiling attenuation class*, which measures how much sound passes through a ceiling panel.

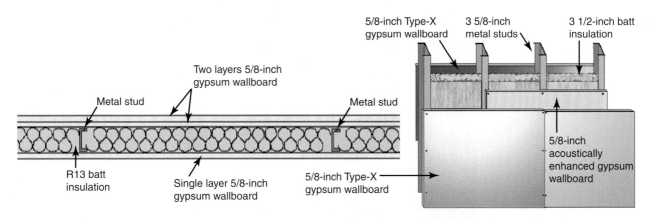

FIGURE 6.30 A partition with an STC rating of 54 is run from slab to slab and is constructed with two layers of ⅝-inch gypsum wallboard on one side with ⅝-inch gypsum wallboard on the other side of three ⅝-inch metal studs with sound attenuation insulation inside the stud cavity.

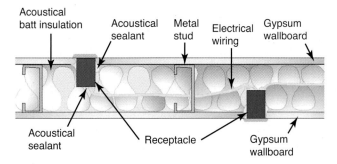

Acoustical batt insulation · Acoustical sealant · Metal stud · Electrical wiring · Gypsum wallboard · Acoustical sealant · Receptacle · Gypsum wallboard

FIGURE 6.31 This section drawing shows the alternating placement of electrical junction boxes to control sound leaks between shared walls.

Plumbing Fixtures and Faucets

Specifying plumbing fixtures like lavatories, toilets, and urinals for a commercial building pose different challenges than those used in residential occupancies. Chapter 5 discussed plumbing codes relative to the required number of water closets and lavatories for occupancy classifications and occupancy loads of a building, but there are other codes that must be followed. Universal accessibility codes regulate how toilet rooms are designed to ensure all persons have complete use of these facilities. Handicap-accessible toilet rooms are mandatory for all occupancy classifications according to the Americans with Disabilities Act passed in 1990. In addition, public toilet rooms must include soap dispensers, paper towel holders, or hand dryers within reach of persons confined to wheelchairs (Figure 6.32).

Sanitation codes require that interior finish materials installed in toilet rooms maintain hygienic conditions. Nonabsorbent wall-surfacing materials like ceramic tile must extend 4 feet AFF, and floors must be installed with nonslip and nonabsorbent materials. Moreover, sanitation codes require that all fixtures be made from nonabsorbent materials. Manufacturers of commercial plumbing fixtures offer toilets, urinals, and lavatories made from porcelain or stainless steel (Figure 6.33).

In addition to sanitation codes, energy codes impose water conservation measures. Toilets must flush with less than 1.6 gallons of

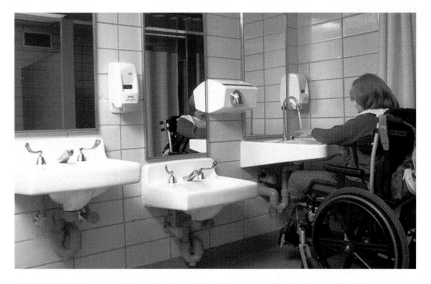

FIGURE 6.32 A woman in a wheelchair has full access to the lavatory, soap dispenser, and hand dryer in this public restroom.

FIGURE 6.33 A row of urinals lines the wall in this public men's room.

Accessibility for All

Universal access codes regulate the design of toilet rooms to ensure persons with physical challenges are able to use the facilities without difficulty. All toilet rooms must be located along accessible routes (those wide enough for a wheelchair to pass—in most cases, 44 inches), and, when inside the toilet room, a 5-foot turning radius must be provided. Toilets are mounted higher than normal for wheelchair access and measure from 17 to 19 inches AFF, with grab bars mounted between 33 inches and 36 inches AFF. Tissue and towel dispensers must be mounted in areas clear of obstructions and within easy reach, and mirrors are mounted lower than usual or fixed with a pivoting device for adjustment. Lavatories are mounted a minimum of 27 inches AFF, with clear space underneath for wheelchair access, and are equipped with either lever-type faucets or motion sensors. For more diagrams and measurements, visit the American with Disabilities website at www.ada.gov.

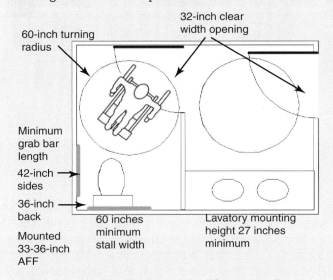

60-inch turning radius

32-inch clear width opening

Minimum grab bar length

42-inch sides

36-inch back

Mounted 33-36-inch AFF

60 inches minimum stall width

Lavatory mounting height 27 inches minimum

Handicap-accessible toilet rooms must include a 5-foot turning radius for maneuvering wheelchairs.

This stall inside a public restroom meets universal access codes with its 5-foot turning radius and stainless steel grab bars.

water, and faucets must be installed with automatic shutoff valves or motion sensors. Sensors save water because they are motion activated and shut off when a person removes his or her hands from underneath the faucet (Figure 6.34).

Wayfinding

During the final planning stages for the interior finish-out phase of construction, the architect or interior designer prepares a *wayfinding* plan as part of the construction documents package. This plan is used to identify the placement of directional signage locating offices, toilet rooms, elevators, fire exit stairwells, areas of refuge, and means of egress. A wayfinding plan includes a legend and a schedule that details the exact specifications for installing the final signage. Visual indicators like color coding helps persons unfamiliar with a building to identify locations for specific areas or departments (Figure 6.35).

Signage and numbering systems locating rooms and offices should be placed adjacent to the latch side of doors, at corridor crossings, and near elevator banks to provide information where directional decisions are normally made. Signs located in the entry and lobby areas, including elevator banks, should use "you are here" maps that show where a person is in relation to areas of egress (like those appearing on the back of the door inside hotel rooms). Signage must be clear enough for people who are unfamiliar with the building to see, understand, and guide them to their destination.

FIGURE 6.34 A touchless faucet is activated by a motion sensor that turns on the water when someone is in front of the sink and shuts off the water when he or she steps away.

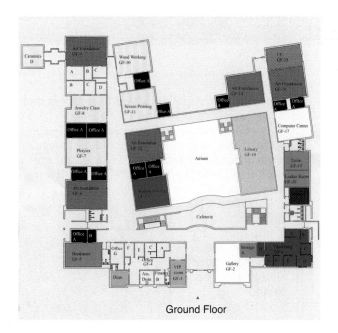
Ground Floor

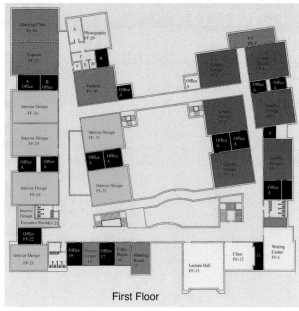
First Floor

FIGURE 6.35 A wayfinding plan is used to indicate locations in a building and is the basis for developing signage in the finished space.

FIGURE 6.36 This sign includes universal pictograms and Braille.

Signage is regulated by both fire safety codes and universal codes. Accessibility codes require that signage contain tactile surfaces including Braille, and regulate mounting heights (Figure 6.36). Signs are installed at 60 inches AFF to the centerline of the sign. International symbols of recognition like pictograms are commonly used to direct persons to the nearest exit. Fire safety codes require that illuminated exit signs be placed at each door leading to exit access, and inside corridors directing persons to the nearest point of exit access (Figures 6.37 and 6.38).

FIGURE 6.37 A directional sign uses a pictogram to indicate where to go to find the nearest exit.

FIGURE 6.38 An illuminated exit sign is mounted to the ceiling where two corridors meet. Notice the manual fire alarm and fire reel mounted to the wall on the right.

LEARN More

Wayfinding for the Visually Impaired

If you've ever been lost trying to find your way through a medical building, looking for the lab or X-ray department, and wander through the seemingly endless maze of corridors, imagine what a person with visual impairments must experience. Persons with visual impairments are extensively trained on how to navigate through both indoor spaces and the outdoors. Visually impaired persons are trained to use a cane to feel the path in front of them and are taught how to pick up clues that orient them to a specific location. Considerable concentration is spent on determining what sound reverberations mean, and let the person know whether the room is large, small, tall, or wide. Blind persons depend on the cane to feel their way through corridors and rooms. Universal access codes require that no object protrude from the wall more than 4 inches. This ensures a person using a cane for navigation will not harm him- or herself by colliding with a drinking fountain or wall-mounted fire extinguisher. Furthermore, tactile flooring surfaces might indicate a change from one area to another—for example, a hallway with resilient flooring changes to carpeting at the threshold of a private office. In addition, signage with Braille augments these navigation clues. The Institute for Innovative Blind Navigation is dedicated to the study, promotion, and development of wayfinding techniques for visually impaired persons to improve safety conditions. For more information, visit their website at www.wayfinding.net.

Informative Websites

Americans with Disabilities Act: www.ada.gov

Armstrong: www.armstrong.com

Innovative Blind Navigation: www.wayfinding.net

National Terrazzo and Mosaic Association: www.ntma.com

U.S. Green Building Council: www.usgbc.org

A

Above finished floor (AFF) The measurement of an object above the finished floor. AFF is standard abbreviation for architectural drawings.

Acoustics The properties of sound, including its transmission and effects.

Aggregate A combination of loose pebbles and sand often used in the making of cement.

Air handler A unit for air-conditioning systems that includes a fan blower, coils, and filter.

Americans with Disabilities Act (ADA) Passed in 1990, this law protects the rights of persons with physical challenges by mandating accessibility requirements.

Ampere The rate at which electricity is delivered to an appliance, limited by the diameter of the wire through which it must flow. Watts divided by volts equal amperes.

Arcuate A structural support method achieved through a series of arches or vaults.

Area of refuge A specially equipped room for persons to await rescue during a fire emergency or in case of building evacuation.

Assembly, assemblies Put together from various parts.

Asymmetric digital subscriber line (ADSL) Communications technology that transfers data over lines faster than traditional copper wiring.

B

Baluster Vertical posts supporting the handrail on a staircase. Also called *banisters*.

Base cabinet A floor-mounted cabinet.

Batt insulation Thick fibers glued to a paper liner and cut into 16-inch-wide sheets for fitting between studs or joints for insulation or sound control.

Beam Any horizontal structural supporting member.

Blue board Gypsum wallboard with blue paper sheeting used to enclose wall studs and ceiling joists.

Boiler Equipment designed to heat water or steam for use in hot-water systems or heating systems.

Branch circuit Distributes electric power to the interior spaces of buildings. Sized according to the amount of current it must carry to sustain a load.

British thermal units (BTUs) Measures thermal output. A unit of heat equal to the amount of heat required to raise 1 pound of water 1°F at 1 atmosphere pressure.

Broadloom carpet Carpeting manufactured on a continuous roll. Also referred to as *sheet carpeting*.

Building codes Requirements set by building officials and code agencies to ensure minimum standards of safety are observed in building construction.

Building permit Authorization from a code official or municipality to allow construction to begin on a project.

Buttressing An external support for a wall.

C

Caissons Foundation piers extending through the earth that rest on sturdy soil or rock.

Cantilevered Any part of a structure extending beyond the point of support.

Carpet tack A thin strip with sharp hooks secured to the perimeter of a room where broadloom carpet will be attached.

Ceiling attenuation class A rating designating the ability of a ceiling material to block airborne sound.

Cement A binding material of limestone (calcite), sand (silica), shale, clay, and iron ore mixed with other aggregates to form concrete.

Cement board A sturdy board used as an underlayment for setting tile.

Central core A concrete shaft that extends vertically through a building and is built to house fire exit stairwells, elevators, and mechanical shafts.

Chiller Equipment used to produce cold water for cooling the interior of a building, including a compressor, condenser, and evaporator.

Chimney A vertical structure running from the firebox of a fireplace up through the roofline for smoke to escape.

Circuit A closed path through which electrical current flows. Also a closed path for running hot water for radiant heating.

Circuit breaker A device that breaks the electrical circuit when it becomes overloaded. The circuit breaker interrupts the electric current that runs through wires to avoid burnout or the risk of fire.

Coaxial cable A cable fitted with copper conducting wires for transmission of low-voltage electric current.

Column Any vertical structural supporting member. Also called a *post*.

Combustible Any material that will burn when ignited.

Compressive strength The ability of a material to resist crushing from exertion of external forces.

Compressor Equipment used in heating and cooling systems to provide enough pressure to pump fluids through a condenser.

Concrete A mixture of cement, aggregate, and water. When cured, the material obtains a rocklike hardness.

Concrete masonry unit (CMU) A concrete block used as building material in masonry construction.

Condenser An exchanger that transfers heat from a compressor to the outside; often water or air cooled.

Conductor, Conductors An electrical component that conducts and confines the flow of electric current within itself. Copper is most widely used and takes the form of wire or cable.

Conduit A pipe or channel for running water, electrical wiring, or communications cabling.

Construction documents A set of drawings and schedules that explain the requirements for construction.

Conveyor systems Systems, including elevators and escalators, that move people through space.

Core drilling Drilling through the concrete slab floor to access electrical conduits from the plenum of the floor below.

Cornice molding A molding placed where the wall meets the ceiling. Also called *crown molding*.

Courses, coursings Rows of brick or stone laid one layer on top of the other. Often used to build a wall.

D

Damper A hinged flap used to control airflow in a chimney or HVAC duct.

Data port A receptacle for the distribution of telephone, Internet, or television cables.

Dead load The weight of the structural elements of a building, such as columns, beams, and slabs. Also the weight of fixed building equipment.

Deflect, deflected The degree to which a beam or lintel bends when weight is applied.

Door handing The location of the hinges on a door and the direction of the swing when opened.

Drywall A gypsum product formed into rigid panels used to cover interior wall studs. Also called *sheetrock*.

E

E-coatings A thin, metallic film applied to windows to reduce the transfer of heat through the glass.

Egress The act of exiting; a means of exit.

Electric current The flow of electricity in an electrical circuit.

Elevation drawing A two-dimensional drawing that shows the vertical plane of a building or interior.

Engineered wood Wood made from waste by-products, including wood fibers, chips, veneers, or dust, held in place with adhesives.

Environmental Protection Agency (EPA) A federal agency charged with the responsibility of reducing pollution and protecting the environment through legislation.

Evaporator Equipment designed to vaporize refrigerant.

Exit access Any path of travel through a corridor, stairwell, room, and so forth, that leads occupants from any space in the building to an exit or area of refuge.

Expansion gap A space between two elements that allows for the expansion of a material as a result of thermal changes.

Extruded steel Molten steel forced through a die that gives the material its shape.

F

FF&E schedule A listing of all furniture, fixtures, and equipment.

Fiber optic cable A cable that uses light to transmit information over glass fibers.

Fiberglass Glass fibers used to reinforce plastics.

Finish carpentry Wood finishes that include doors, stairs, paneling, moldings, and trims.

Fire annunciation Equipment designed to warn of potential fire or emergency situations using high-decibel alarms and flashing lights.

Fire rating The amount of time expressed in hours that a material will withstand damage from exposure to fire.

Fire resistive Any material that slowly resists damage from fire. A material that resists burning.

Firebox The part of a fireplace or stove where the fire burns.

Flashing A flat piece of metal installed between the roofing material and joints or seams for weatherproofing.

Flitch Sliced segments from a log that are bundled together in a batch for veneers. A series of veneers cut from the same log.

Float glass A type of glass formed by pouring molten glass over tin. The glass "floats" over the tin while cooling, forming a solid sheet.

Floor joist A system of support beams that hold up a finished floor.

Floor plan, floor plans A two-dimensional drawing used in construction that shows the floor of a building or structure.

Flue A vertical lining made of noncombustible materials inside a chimney for release of smoke and embers from a fire to the outside air.

Footing A pad or block, usually concrete, that acts as a base for foundation systems.

Footprint The perimeter plan of a house, building, or structure. It outlines the overall area of the structure.

Formaldehyde An organic compound used in polymers and adhesives.

Formwork A temporary platform or mold used for poured-concrete construction. A temporary support structure usually used in site-cast concrete work.

Foundation The main support system for a building or structure usually placed on or below the ground.

Foundation plan A two-dimensional drawing that shows the type of foundation system for construction.

Framing plan A two-dimensional drawing that shows how a structure is framed for structural integrity.

Furring channel A narrow strip of metal attached to concrete or concrete block for securing a finish material to its surface.

G

Gable, gables The face end of a steeply pitched roof.

Galvanized steel Steel with an anticorrosive finish of zinc.

General contractor A licensed individual who oversees the construction of a building.

Geothermal Heat from the earth. Used in producing sustainable energy.

Glazed partitions Walls made from glass or having portions of glass in the assembly.

Glazed tile A glossy finish applied to tiles before firing, yielding water-repellant tendencies.

Glazing The glass in a wall, door, or window.

Grade Ground level.

Gray water Waste water from sinks, showers, tubs, and washing machines that is recycled and used for irrigation.

Green design Design and construction practices intended to save natural resources and protect the environment.

Ground fault circuit interrupter (GFCI) An electrical device that stops the flow of electricity when the circuit is exposed to water or moisture.

Grout A mortar-type substance used to fill the spaces between set tiles.

Gypsum Calcium sulfate used in the production of wallboard, plaster, and stucco.

Gypsum plaster Plaster made with lime, sand, and water.

Gypsum wallboard A wall-enclosing material made from sandwiched layers of a gypsum core and heavy paper. Sometimes called *drywall*.

H

HDF See *High-density fiberboard*.

Header A horizontal support used in the framing of doors and windows.

Hearth The area immediately in front of the firebox of a fireplace. According to code requirements, hearth materials must be noncombustible.

Heat gain/heat loss The amount of heat infiltrating in to or out of a building.

Heating, ventilation, and air-conditioning (HVAC) systems Equipment designed to maintain thermal comfort within a building or structure.

High-density fiberboard (HDF) A high-density engineered material comprised of wood particles, resins, and adhesives formed into sheet panels.

High-rise A building with a height more than 35 meters or 114 feet that requires mechanical transport devices to reach the upper floors.

Hoist, hoisting A device used to raise or lift weight.

Hypostyle hall A room in which columns are used to support the roof or ceiling.

I

I beams Extruded steel used as a structurally stable construction material, the shape of which forms the letter "I."

Indoor air quality The measurement of the purity of indoor air.

International Code Council (ICC) A membership association dedicated to ensuring building safety and fire prevention, and developing a system of codes used to construct residential and commercial buildings.

International energy conservation code (IECC) Requirements recommended and published by the International Code Council to ensure energy-saving and energy conservation devices are incorporated into building construction.

International plumbing code (IPC) Requirements recommended and published by the International Code Council to ensure minimum standards of sanitation are observed in building construction.

J

Jamb A vertical frame located on the sides of windows or doors.

Joint compound A moist paste used to conceal joints or holes. Also called by the brand name *Spackle*.

Joist A horizontal structural support device usually used to support floors and ceilings.

Junction box Wiring is brought into these boxes, which are spliced with switches or receptacle outlets held together with wire nuts. Shapes are square for receptacles and switches, and hexagonal for ceiling outlets. Material may be metal or plastic.

K

Keystone The centermost block in an arch that keeps all components intact.

Kilowatt The measurement of consumed electricity in 1,000 watts of power.

L

Laminate, laminating The bonding together of two or more materials.

Laminated glass Layered glass, or the layering of glass with other materials to create safety glass, insulated glass, or glass with specific properties.

LAN (local area network) A system linking computers, printers, and other equipment to one main source to share technology.

Lateral forces Loads bearing internally or externally on a structure in a manner parallel to the ground. Usually caused by wind, seismic changes, or floodwaters.

Lavatory, lavatories A sink used for hand washing.

LCD glass Liquid crystal display glass. A material that changes from transparent to opaque with electric current.

Leverage A mechanism for exerting force to move an object.

Linoleum A flooring material made from linseed oil, wood pulp, and resin.

Lintel A horizontal structural member that supports weight from above. A lintel provides the overhead support for an opening such as a door, portal, or window.

Live load The weight of movable items in a building, such as furniture, people, and equipment.

Load The force of weight placed upon a structural element.

Load bearing A structural element designed to resist or bear the weight of other objects or materials.

Low E-coatings Low-emittance coatings applied to glass that limit UV rays passing through.

Low-rise A building with at least one level above the ground floor.

M

Mantel The finish surrounding a fireplace, including an overhead shelf.

Masonry Brick, tile, stone, or concrete materials used in construction.

Matte glaze A low-gloss finish applied to tiles before firing that yields water-repelling tendencies.

MDF See *Medium-density fiberboard*.

Mechanical systems Environmental systems within a building such as air-conditioning, heating, and plumbing.

Medium-density fiberboard (MDF) A medium-density engineered material comprised of wood particles, resins, and adhesives formed into sheet panels.

Megalith Large stones used in construction during ancient times.

Metal decking Corrugated sheets of metal used in floor and roof systems for commercial structures.

Metal studs Extruded steel framing members used in wall or ceiling construction.

Mid-rise A building ranging from five to 12 stories in height.

Millwork Woodwork carpentry including stair construction, door and window trims, baseboards, and moldings. Wood materials milled in a plant.

Molding profile The shape of millwork trim as seen from the side.

Mortar, mortar bond A bonding agent consisting of cement, sand, and lime used to set concrete, stone, or brick.

Muntins Thin structural bars that hold the glass in place in a window.

N

National Electric Code (NEC) Requirements recommended and published as part of the National Fire Protection Agency to ensure minimum standards for electrical safety are incorporated into building construction.

National Fire Protection Agency (NFPA) An association that develops, publishes, and disseminates consensus codes and standards intended to minimize the possibility and effects of fire and other risks in building construction.

Nave The central area within a church that extends from the entry toward the apse.

Noise reduction coefficient (NRC) A rating that measures the ability of a material to absorb sound.

Noncombustible Any material that will not burn or ignite.

O

Occupancy classifications The basis for applying building codes according to the type of user-based function within a building. The main classifications include Assembly, Business, Education, Institutional, Mercantile, and Residential.

Occupancy load The number of people occupying a space according to a formula based on square footage and occupancy classification.

Oculus A round opening in the top of a domed structure.

Open-web steel joist Load-bearing structural member made from steel and used to support floors and roofs.

Oriented strand board (OSB) An engineered material comprised of shaped-wood strands that are cross-layered and formed into sheets.

P

Parapet A low wall placed along the edge of a flat roof.

Passive solar Maximizing the sun's energy to keep a building or structure warm or cool.

Performance codes Building codes that measure the performance of precautionary measures taken to protect the health, safety, and welfare of building occupants.

Photovoltaic A device used to capture light from the sun and turn it into electricity.

Pilings Foundation piers extending through the earth that rest on sturdy soil or rock and are made of concrete-filled steel piping.

Plaster A mixture of gypsum or limestone, water, and fine aggregate like sand used as an interior finishing material.

Plenum The space between the finished ceiling and the slab of the floor above.

Plywood An engineered material comprised of laminating thin veneers of wood that are cross-layered and formed into sheets.

Polyvinyl chloride (PVC) A man made material made from chlorine and petroleum based carbon to produce a wide range of plastics.

Post A vertical support designed to carry weight.

Postproduction waste Leftover matter or materials from the production of manufactured products.

Posttension cabling Steel cables placed in tension by stretching. Posttension cables are used to reinforce concrete.

Potable Water purified for drinking.

Pounds per square inch A unit that measures the amount of pressure placed on a structural support. Also called *PSI*.

Pozzolana Volcanic ash prevalent in southern Italy comprised of silica and aluminum oxide, and used as a form of cement.

Prefabricated Any item built at a factory or away from the construction site, such as steel beams.

Prescriptive codes Building codes with specific instructions on how to protect the health, safety, and welfare of building occupants.

Pre-stressed concrete A prefabricated concrete structural member in which a steel cable is placed under tension as concrete is poured into formwork. Pre-stressing adds tensile strength to the inherent compressive strength of concrete.

Primer A coating applied to bare wood or wallboard before finish coats are applied.

Pulley A circular device for lifting weight. A grooved wheel is fitted with a rope and, when pulled, it lifts a weight attached to the opposite end of the rope.

R

Rafter A wood timber or engineered wood beam used to support a roof.

Rails Horizontal support members for paneling, doors, windows, and staircases.

Raised access floor A subfloor installed above the concrete slab floor that leaves space for electrical or data cabling. The subfloor is fitted with access panels to reach the cabling underneath.

Raised paneling Section of wood paneling with beveled edges.

Rebar A steel material either in mesh form or narrow bars used in poured concrete for reinforcement.

Receptacle A wiring device installed with an outlet box for the connection of an electrical apparatus through an attachment plug.

Reengineered The redesign of materials or products to achieve better efficiency in performance or to attain a lower ecological impact.

Reflected ceiling plan A plan of the ceiling as seen from inside the space. This plan shows locations of light fixtures, ceiling articulation, and switching.

Reinforced concrete Concrete poured over steel rebar for added strength.

Renovate The process of updating or changing an existing building or structure.

Restoration The process of returning a building or structure to its original condition.

R factor A rating system used to determine a material's ability to resist heat transfer.

Riser The part of a staircase creating the vertical rise of the step.

Rough opening dimensions The measurements of a clear and unobstructed opening for fitting in doors, windows, cabinetry, or appliances.

Rough-in A stage of installation before final finishing or enclosure.

Rubble stone Irregular-shaped stones used in construction.

Runner track A channel anchored to the subfloor or floor joists to secure supports for a wall.

S

Safety glass A type of glass that breaks into small pieces without sharp edges.

Sandblast Ejecting sand onto a surface using high pressure to remove debris or to etch its surface.

Scaffolding Temporary platforms built on a construction site for workers to reach high places.

Section drawing A two-dimensional drawing that shows the vertical plane of a building or interior as it is cut through the thickness of the walls, roof, ceiling, floors, and slab.

Sewer gases Noxious fumes emitted from decomposing waste.

Sheathing A material used as a protective covering.

Sheetrock See *drywall*.

Short circuit When an abnormally high current flows through the circuit.

Site-cast concrete Concrete poured into formwork on the construction site; not prefabricated.

Soil stack A plumbing pipe that removes human waste from the toilet into a septic tank or sewer system.

Solar heat gain coefficient A rating that measures the amount of heat transferred through a material.

Sound transmission The ability of sound to travel through elements like walls, ceilings, floors, or doors in a building.

Span The distance measured between two columns or posts.

Splay, splayed Columns, posts, or any other vertical means of support positioned at a slight angle.

Stiles Vertical support members for paneling, doors, windows, and staircases.

Stock cabinets Premade and readily available cabinetry.

Stock moldings Premade and readily available moldings.

Stock plans A set of construction documents predesigned and available for purchase.

Stringer The structural sloping side member of a staircase that supports risers and treads.

Strip flooring A flooring material made up of narrow sections of wood held together with tongue-and-groove joints.

Stucco A mixture of gypsum or limestone, water, cement, and fine aggregate like sand used as an exterior finishing material.

Stud A slender support member used to frame a wall or ceiling. Often made from wood or steel.

Subfloor The area beneath the finished floor. Usually comprised of plywood or OSB sheathing secured to floor joists.

Substrate material The element that provides dimensional stability for the application of a finish material, such as gypsum wallboard for paint, or OSB sheathing for roofing tiles.

Substructure The part of a building below ground.

Superstructure The part of a building above ground.

Suppression systems Devices that work toward controlling, containing, or extinguishing smoke and fire.

Sustainable Any natural substance or material that replenishes itself on a regular basis, such as grasses or wool.

Switch, Switches A device for making, breaking, or changing the connections in an electrical circuit.

T

Tape and bedding The process of covering gaps between installed gypsum wallboard to give walls and ceilings a seamless and smooth surface.

Tempered glass High-strength glass that breaks into small pebblelike pieces without sharp edges.

Tensile strength The ability of a material to withstand stretching, bending, or elongation from the pressure of external forces.

Thatch Straw, reed, or grasses bundled together and used as a building material.

Thermostat A mechanical device used to measure and maintain temperature.

Timber frame A structural support system for a building or structure comprised of wooden timbers or studs.

Tongue-and-groove A type of joint used to interlock two pieces into one whereby a flange in one piece fits into a recess on the other piece.

Trabeated A structural support system that relies on beams and columns.

Tread The part of a staircase where one places his or her foot to climb.

Truss, trusses A structural support beam that maximizes the stability of a triangular force.

U

Underlayment Any material applied before the finishing material, such as tar paper for roofing or cement board for ceramic tile.

Unglazed tile Tiles that have been fired without the addition of pigments or sheen.

Urethane A clear polymer compound.

User group A category used to define a building's function.

V

Valve A device used to control the flow of water or air through a pipe.

Variable air volume (VAV) Mechanical equipment used in heating and air-conditioning systems to mix heated and cooled air to maintain thermal control set by a thermostat.

Vault An overhead ceiling or structural system based on arches.

VAV box Variable air volume control box that controls airflow and regulates air temperature for zoned areas in a building.

Veneer Thin layer of any material applied to a substrate for dimensional stability. Examples include plastic laminate, wood, metal, or stone.

Vent stacks Plumbing pipes that run vertically through the roof of a building from the drain. Used to remove toxic sewer gases from an interior space.

Ventilation The circulation of fresh air and removal of stale air throughout a building.

Vitreous china Ceramic with less than a 0.5% absorption rate.

Volatile organic compounds (VOCs) Chemical compounds that emit gases that may act as carcinogens.

Volt, voltage The measured rate of electrical pressure traveling through wiring, either 120 volts or 240 volts.

W

Wainscoting A special wall treatment applied to the lower portion of a wall surface.

Wallboard Rigid sheets of any material that encloses the stud walls, often gypsum wallboard.

Waste stack A plumbing pipe that removes waste water from sinks, tubs, showers, or washing machines into a septic tank or sewer system.

Water closet A toilet or toilet room.

Wayfinding A method of human navigation inside a building. Includes signage such as directories, room names or numbers, or maps.

Wet wall A wall fitted with plumbing supply pipes.

Wood plank flooring A flooring material made up of wide sections of wood held together with tongue-and-groove joints.

Woodworking The process of making things from wood by cutting, sawing, carving, or milling.

Y

Yurt A circular and often domed shelter, usually temporary.

Aghayere, Abi O. and Jason Vigil. *Structural Steel Design: A Practice Oriented Approach.* Upper Saddle River, NJ: Prentice Hall, 2008.

Althouse, Andrew D., Carl H. Turnquist, and Alfred F. Bracciano. *Modern Refrigeration and Air Conditioning Principles.* Tinley Park, IL: Goodheart-Wilcox, 2004.

American Wind Energy Association. *Market Update: Record 2009 Leads to Slow Start in 2010.* Accessed January 20, 2011 online at www.awea.org/documents/factsheets/Market_Update_Factsheet.pdf

Andres, Cameron K. and Ronald C. Smith. *Principles and Practices of Commercial Construction,* 8th ed. Upper Saddle River, NJ: Prentice Hall, 2008.

Baham, Reyner. *The New Brutalism.* Architectural Press, London, 1966.

Dagostino, Frank R. and Joseph B. Wujek. *Mechanical and Electrical Systems in Architecture, Engineering and Construction,* 5th ed. Upper Saddle River, NJ: Prentice Hall, 2009.

Gardner, Larry and Leo Meyer. *Tips for Residential HVAC Installation.* Carol Markos, Ed. Steve Meyer, Illus. Hayward, CA: LAMA Books, 2006.

Gipe, Paul. *Wind Power: Renewable Energy for Home, Farm, and Business.* White River Junction, VT: Chelsea Green Publishing, 2004.

Haines, Roger and C. Lewis Wilson. *HVAC Systems Design Handbook*, 4th ed. New York: McGraw-Hill, 1998.

International Code Council. *Smoke Control Provisions of the 2000 IBC: An Applications Guide.* Washington, DC: International Code Council, 2002.

Jensen, Rolf. *Designing and Building with the International Building Code*, 2nd ed. Kingston, MA: R. S. Means, 2003.

Joyce, Michael A. *Blueprint Reading and Drafting for Plumbers*, 2nd ed. Florence, KY: Delmar Cengage Learning, 2009.

Kachadorian, James. *The Passive Solar House: The Complete Guide to Heating and Cooling Your Home,* 2nd ed. White River Junction, VT: Chelsea Green Publishing, 2006.

Kaufman, Harry F. *A Structures Primer.* Upper Saddle River, NJ: Prentice Hall, 2008.

Kreh, Dick. *Building with Masonry: Brick, Block and Concrete.* Newtown, CT: Taunton Press, 2002.

Kubba, Sam. *Blueprint Reading: Construction Drawings for the Building Trades.* New York: McGraw-Hill Professional, 2009.

Lechner, Norbert. *Heating, Cooling, Lighting: Design Methods for Architects,* 3rd ed. Hoboken, NJ: Wiley, 2008.

Massey, Howard C. *Basic Plumbing.* Carlsbad, CA: Craftsman Book, 1994.

Massey, Howard C. *Planning Drain, Waste and Vent Systems.* Canoga Park, CA: Builder's Book. 2002.

McCormac, Jack C. and Russell Brown. *Design of Reinforced Concrete,* 8th ed. Hoboken, NJ: Wiley, 2009.

Miller, Mark R. and Rex Miller. *Miller's Guide to Foundations and Sitework.* New York: McGraw-Hill Professional, 2005.

Miller, Mark, Rex Miller, and Glenn Baker. *Miller's Guide to Home Plumbing.* New York: McGraw-Hill Professional, 2005.

"NCIDQ. Definition of Interior Design: NCIDQ." *Home.* Web. 20 Dec. 2010. www.ncidq.org/AboutUs/AboutInteriorDesign/DefinitionofInteriorDesign.aspx

Portland Cement Association. *Concrete Systems for Homes and Low-Rise Construction.* New York: McGraw-Hill Professional, 2006.

Setareh, Mehdi and Robert M. Darvas. *Concrete Structures.* Upper Saddle River, NJ: Prentice Hall, 2007.

Smith, Ronald C., Ted L. Honkala, and W. Malcolm Sharp. *Principles and Practices of Light Construction*, 6th ed. Upper Saddle River, NJ: Prentice Hall, 2003.

Spence, William P. *Construction Materials, Methods, and Techniques,* 2nd ed. Florence, KY: Delmar Cengage Learning, 2007.

"The Construction of the RWE Tower." *Space Modulator* 86 (1999): 81–90. Web. 20 Dec. 2010.

U.S. Deptartment of Energy. *20% Wind Energy by 2030: Increasing Wind Energy's Contribution to U.S. Electricity Supply.* Washington, DC: U.S. Government, 2008.

Vedavarz, Ali, Sunil Kumar, and Muhammed Hussain. *HVAC: Heating, Ventilation & Air Conditioning Handbook for Design & Implementation.* New York: Industrial Press, 2007.

Wahl, Iver. *Building Anatomy: An Illustrated Guide to Why Structures Work.* New York: McGraw-Hill Professional, 2007.

Weidhaas, Ernest R. *Reading Architectural Plans for Residential and Commercial Construction*, 5th ed. Upper Saddle River, NJ: Prentice Hall, 2001.

Wyatt, David J. and Hans W. Meier. *Construction Specifications: Principles and Applications.* Florence, KY: Delmar Cengage Learning, 2008.

Zachariason, Rob. *Blueprint Reading for Electricians,* 3rd ed. Florence, KY: Delmar Cengage Learning, 2009.